Water Quality

Water Quality
Processes and Policy

Edited by

Stephen T. Trudgill
Department of Geography, University of Cambridge, UK

Des E. Walling
Department of Geography, University of Exeter, UK

and

Bruce W. Webb
Department of Geography, University of Exeter, UK

JOHN WILEY & SONS, LTD
Chichester • New York • Weinheim • Brisbane • Singapore • Toronto

Other Wiley Editorial Offices

John Wiley & Sons, Inc., 605 Third Avenue,
New York, NY 10158-0012, USA

WILEY-VCH Verlag GmbH, Pappelallee 3,
D-69469 Weinheim, Germany

Jacaranda Wiley Ltd, 33 Park Road, Milton,
Queensland 4064, Australia

John Wiley & Sons (Asia) Pte Ltd, Clementi Loop #02-01,
Jin Xing Distripark, Singapore 129809

John Wiley & Sons (Canada) Ltd, 22 Worcester Road,
Rexdale, Ontario M9W 1L1, Canada

Library of Congress Cataloging-in-Publication Data

Water quality: processes and policy / edited by Stephen T. Trudgill,
Des E. Walling, and Bruce W. Webb.
p. cm.
Includes bibliographical references and index.
ISBN 0–471–98547–3 (alk. paper)
1. Water quality. 2. Water—Pollution. 3. Water quality
management—Government policy. I. Trudgill, Stephen T. (Stephen
Thomas), 1947– . II. Walling, D. E. III. Webb, Bruce.
TD370.W593 1999
363.739′456—dc21   99–10013
CIP

British Library Cataloguing in Publication Data

A catalogue record for this book is available from the British Library

ISBN 0 471 98547 3

Typeset in 10/12pt Times by Vision Typesetting, Manchester
Printed and bound in Great Britain by Biddles Ltd, Guildford and King's Lynn
This book is printed on acid-free paper responsibly manufactured from sustainable forestry, for which at least two trees are planted for each one used for paper production.

Contents

List of Contributors

Steven Anthony, ADAS Research and Development, Wergs Road, Wolverhampton, WV6 8TQ, UK.

Adrian Armstrong, ADAS Land Research Centre, Gleadthorpe, Meden Vale, Mansfield, Notts, NG20 9PF, UK.

David Botterill, School of Leisure and Tourism, Colchester Campus, University of Wales Institute, Cardiff, CF61 1SS, Wales, UK.

Tim Burt, Department of Geography, University of Durham, Science Laboratories, South Road, Durham, DH1 3LE, UK.

Dan Butterfield, Aquatic Environments Research Centre, Department of Geography, University of Reading, Reading, RG6 6AB, UK.

John Catt, IACR Rothamsted, Harpenden, Herts, AL5 2JQ, UK.

Meng Fan Chen, Department of Geography, University of Aberdeen, Aberdeen, Scotland, AB24 3UF, UK.

Bob Crabtree, WRc plc, Frankland Road, Blagrove, Swindon, SN5 8YF, UK.

John Crowther, Department of Geography, University of Wales, Lampeter, Ceredigion, SA48 7ED, UK.

Bryan Ellis, Urban Pollution Research Centre, Middlesex University, Queensway, Enfield, EN3 4SF, UK.

Lorna Fewtrell, CREH at IGEF, University of Wales, Aberystwyth, Ceredigion, SA23 3DB, UK.

Jay Fleisher, Department of Preventive Medicine, College of Medicine, Health Sciences Building, State University of New York, Brooklyn, NY, USA.

Ian Foster, Geography Department, Coventry University, Priory Street, Coventry, CV1 5FB, UK.

Colin Green, Flood Hazard Research Centre, Middlesex University, Queensway, Enfield, EN3 4SF, UK.

Mark Hancock, Geography Department, Coventry University, Priory Street, Coventry, CV1 5FB, UK.

Graham Harris, ADAS Land Research Centre, Gleadthorpe, Meden Vale, Mansfield, Notts, NG20 9PF, UK.

Brian Ilbery, Geography Department, Coventry University, Priory Street, Coventry, CV1 5FB, UK.

Gerry Jackson, Public Services Department, States of Jersey, South Hill, St. Helier, Jersey, JE4 8UY, UK.

Helen Jarvie, Institute of Hydrology, Maclean Building, Crowmarsh Gifford, Wallingford, Oxon, OX10 8BB, UK.

Tim Jarvis, Pesticides Safety Directorate, Mallard House, Kings Pool, 3 Peasholme Green, York, YO1 2PX, UK.

David Kay, CREH at IGEF, University of Wales, Aberystwyth, Ceredigion, SA23 3DB, UK.

Pauline Kneale, School of Geography, University of Leeds, Leeds, LS2 9JT, UK.

Stuart Lane, Department of Geography, University of Cambridge, Downing Place, Cambridge, CB2 3EN, UK.

Eunice Lord, ADAS Research and Development, Wergs Road, Wolverhampton, WV6 8TQ, UK.

R. Malcolm, Department of Geography, University of Aberdeen, Aberdeen, Scotland, AB24 3UF, UK.

Adrian McDonald, School of Geography, University of Leeds, Leeds, LS2 9JT, UK.

Colin Neal, Institute of Hydrology, Maclean Building, Crowmarsh Gifford, Wallingford, Oxon, OX10 8BB, UK.

Cliff Nelson, Coastal Studies Institute, School of Geosciences, University of Sydney, Sydney, NSW 2006, Australia.

Gary O'Neill, Yorkshire Water Services Ltd, George Street, Bradford, BD1 5PZ, UK.

Edwin Ongley, UNEP/WHO – GEMS/Water Programme, Canada Centre for Inland Waters, PO Box 5050, Burlington, Ontario, L7R 4A6, Canada.

Paul Quinn, Water Resource Systems Research Laboratory, Newcastle University, Newcastle upon Tyne, NE1 7RU, UK.

G. Rees, Farnborough College of Technology, Boundary Road, Farnborough, Hampshire, GU14 6SB, UK.

A. Richards, Matra British Aerospace Engineering Dynamics, PC510, Filton, Bristol.

Keith Richards, Department of Geography, University of Cambridge, Downing Place, Cambridge, CB2 3EN, UK.

Alice Robson, Institute of Hydrology, Maclean Building, Crowmarsh Gifford, Wallingford, Oxon, OX10 8BB, UK.

Sudanshu Sinha, Department of Geography, University of Cambridge, Downing Place, Cambridge, CB2 3EN, UK.

David Ingle Smith, Centre for Resource and Environmental Studies, Australian National University, Canberra 0200, ACT, Australia.

Christopher Soulsby, Department of Geography, University of Aberdeen, Aberdeen, AB24 3UF, Scotland, UK.

Steve Trudgill, Department of Geography, University of Cambridge, Downing Place, Cambridge, CB2 3EN, UK.

Des Walling, School of Geography and Archaeology, Department of Geography, University of Exeter, Amory Building, Rennes Drive, Exeter, EX4 4RJ, UK.

Bruce Webb, School of Geography and Archaeology, Department of Geography, University of Exeter, Amory Building, Rennes Drive, Exeter, EX4 4RJ, UK.

Paul Whitehead, Aquatic Environments Research Centre, Department of Geography, University of Reading, Reading, RG6 6AB, UK.

Allan Williams, Faculty of Applied Sciences, Bath Spa University College, Newton Park, Newton St Loe, Bath, BA2 9BN, UK.

E. Wilson, National Power, Windmill Business Park, Swindon, Wiltshire, SN5 6PV, UK.

Shuang-ye Wu, Department of Geography, University of Cambridge, Downing Place, Cambridge, CB2 3EN, UK.

Mark Wyer, CREH at IGEF, University of Wales, Aberystwyth, Ceredigion, SA23 3DB, UK.

Foreword

The Lord Lewis of Newnham
Chair of the Royal Commission of Environmental Pollution,
1986–1992

It has been argued that the present position of the use of water in the world is imposing a problem of the same magnitude as the oil crisis of the 1970s. The managing of water resources is by no means a new problem and the control and use of water has been a matter of major importance from time immemorial, with the rights and use of waters for both domestic and commercial use dominating the course of history from ancient times. It is therefore not surprising that legislation concerning water quality and control has been a major feature of EU legislation involving more than 30 directives. The present projected EU programme on River Basin Management merely reflects the problems of generations of people within the European Union and reflects the difficulties that have taxed countries in controlling and utilising the use of the rivers over the past 2000 years. Thus the problems of the rights of downstream users in any river basin have always been points of contention. It is therefore not surprising that the EU has placed much emphasis in its regulations on water quality and control in an attempt to rationalise the use of this important and essential facility within the community.

The major concern in any environmental considerations, which is particularly applicable to problems associated with water, is that of population growth. This situation is further complicated by the movement of people from rural to urban communities. According to World Bank statistics, by 2030 it is estimated that the world population will have doubled in size, and for the under-developed countries the main growth will be within cities (160%) whilst for rural areas the growth will be much more limited (10%). A further complication that arises is that not only are the numbers of people increasing at this remarkable rate but the demands of modern-day living, with the increasing use of domestic appliances that require water, provide a very significant increase in the water consumption per capita. It has been estimated that over the next four decades the domestic demand for water in the western community will rise by a factor of five. Further to this must be added the extra loading from industry and increasing demand from agriculture. These factors will obviously aggravate any of the problems associated with water conservation and usage.

There is a considerable difference in the problems of water conservation and usage between the developed and developing countries. Water problems in the developing parts of the world are particularly acute; for instance in the countries of Latin America less than 2% of sewage is treated and in Mexico 90% of wastewater treatment plants are not functioning. In Jakarta 66.6% of the population depend upon the use of groundwater for their water supply, and in the northern part of Jakarta the water table has been falling continuously from 1970 owing to over-pumping of the groundwater supply. This has led to contamination of the groundwater in a 5–10 km belt in the coastal plain by seawater incursion, leading to chloride content of the water being up to 500 mg per litre when the World Health Organization guideline is 2 mg per litre. The boiling of the water for purification is estimated to cost US$50 million and leads to both air pollution and deforestation. The introduction of desalination plants would certainly provide a solution that would be financially realistic on a relatively short time scale, but is not being carried out at the moment.

For the developed world the pollution problems of groundwater are also a major problem. In the UK, 30% of domestic water arises from groundwater sources, and for certain parts of the country this factor may rise to 70%. It is therefore of prime importance to maintain the standards of these water sources as once contaminated they may take many years if not centuries to recover. The National Rivers Authority, now the Environment Agency, has paid considerable attention to this problem and has prepared "vulnerability" maps for aquifers throughout the majority of the country, with the creation of protection zones. This has placed strict limits on any potential developments near to aquifers. However, the possibility of air-borne pollution and particularly agricultural spillage of slurry or silage effluent can readily lead to pollution problems not only of surface waters but also to groundwater sources.

The articles in this book cover many of these points, and as with all present-day problems dealing with the environment, the discussions spill over into many disciplines – science, economics, engineering, social science and law. The recognition and detection of the problem in environmental pollution is often based on monitoring and the scientific interpretation of the data. The solution in many instances is dependent on economic considerations. The current practice for the use of life cycle analysis of the problem involves an economic recognition of the value of many environmental factors that are not easy to quantify in monetary terms. This can lead to an ambiguous assessment of the financial liabilities. Clearly for a controlled assessment of environmental problems there must be a legal basis for the recognition and detection as well as the implementation of the rules. Finally, the projected solution to a given problem, the so-called "best practicable environmental option", may provide difficulties in other areas such as employment and economic development for a particular region. In many parts of the developing world, pollution control is viewed with a degree of resistance as the implementation is often seen as involving restrictions within industry, potential unemployment, or increases in commodity prices.

The consideration of water, therefore, provides an excellent vehicle for the general considerations of the broad base for any approach to the study of environmental problems as well as highlighting one of the major concerns in modern society. The present book summarises and makes significant contribution to the particular and the general issues involved with these problems.

Robinson College, Cambridge, 1998

Introduction

Bruce W. Webb, Des E. Walling and Steve T. Trudgill

This volume is the by-product of a one-day meeting on Water Quality: Processes and Policy held during *Exeter '97*, the Royal Geographical Society – Institute of British Geographers (RGS-IBG) Annual Conference which took place at the University of Exeter, Devon, UK, in early January 1997. The session was convened by two of the RGS-IBG Research Groups, namely the British Geomorphological Research Group and the Environmental Research Group. The aim of the meeting was to address the relationship between science and policy making with respect to water quality in river and other environments. The answering of questions such as "do policy makers use scientific underpinning or are their decisions more politically motivated?", and "should scientists be constrained by policy-making contexts?" was one of the objectives of this meeting. The 15 chapters in this book are based on papers given at Exeter and on a number of invited contributions. The book is divided into three sections, which contain chapters dealing with global perspectives, with particular scientific issues informing water quality policies, and with the links between science and policy in the water quality field.

Water quality is a vast and complex topic. It embraces all phases of the hydrological cycle from precipitation inputs, through terrestrial surface and groundwater systems to the marine environment into which freshwater runoff ultimately discharges. Water quality can be defined by numerous physical, chemical and biological parameters, is subject to a wide range of natural controls and human influences, and exhibits complex variations over different spatial and temporal scales (Peters *et al.*, 1997, 1998).

Although not exclusively the result of human impacts, very many water quality problems existing in the world today have arisen through the use of water for domestic, agricultural and industrial purposes. As a consequence, surface, ground and coastal

Water Quality: Processes and Policy. Edited by Stephen T. Trudgill, Des E. Walling and Bruce W. Webb.
© 1999 John Wiley & Sons Ltd.

waters have become polluted with a wide range of contaminants including organic material, nutrients, salts, acidifying compounds, heavy metals, organic micropollutants, radioactivity and sediments (e.g. Meybeck *et al.*, 1989).

GLOBAL PERSPECTIVES

In the Foreword to this book, The Lord Lewis of Newnham highlights the fact that problems of world water use in the 1990s are creating a crisis of similar proportion to that associated with the use of oil in the 1970s. Fuelled by future population growth and increasing urbanisation within less developed countries, and an ever-increasing demand for water in developed nations, especially in the domestic and agricultural sectors, problems of water scarcity and contamination are set to rise well into the twenty-first century.

In his review of the linkage between water quality science and policy at the global scale, Ed Ongley (Chapter 1) makes the case that issues of water quality and pollution are increasingly seen to be as important as issues of water shortage. However, freshwater vulnerability is likely to be most severe when problems of water contamination are combined with those of water scarcity. Ongley speaks of an emerging global crisis of water quality, which manifests itself through an annual death toll of five million people from water-borne diseases, contamination of freshwater and marine ecosystems, loss of biodiversity and pollution of groundwater resources, amongst other ways. He argues that to combat this crisis, water quality monitoring, especially in developing, but also in developed, nations needs to provide more appropriate information for interpreting local ecosystem and public health status and for understanding global problems of biodiversity and atmospheric contamination. In many countries, improvements in monitoring also need to be accompanied by an enhanced political awareness of water quality issues and the modernisation of technical, institutional, legal, training and support aspects of water quality programmes.

SCIENCE FOR POLICY

Reliable estimates of river-borne fluxes of contaminants are crucial to quantifying the sources of pollution in coastal waters and to establishing suitable management strategies and legislative frameworks for controlling water quality problems in coastal environments. While several studies have highlighted the problems associated with the computational methodologies that are often used to derive river loads, Helen Jarvie and her colleagues (Chapter 2) demonstrate that in the lower reaches of major rivers, other factors, such as the location of the sampling point with respect to tidal and saline limits, and timing of sampling in relation to the tidal cycle and river flow conditions, also markedly affect the reliability of river load estimates, especially in the case of suspended sediment and particulate-associated contaminants. In the context of international water quality monitoring programmes, such as that co-ordinated by the Paris Commission for the North Sea region, there are significant national and regional differences in the location of sampling sites and the tidal and river flow conditions under which sampling is undertaken. Jarvie *et al.* believe these may jeopardise comparison of riverine mass loads at the European regional scale.

In the context of pollution of nearshore bathing waters by faecal bacteria, David Kay and his colleagues (Chapter 3) argue that care must be taken to include all potential sources of contamination. Their studies in the States of Jersey and elsewhere have demonstrated that storm-period runoff from catchments draining directly to the coastal zone may be a very important source of faecal bacteria. In consequence, the expensive measures that have been, and are being, taken in the UK to improve the effluent of sewage treatment works discharging to nearshore waters may not be effective in eliminating significant contamination by faecal bacteria. In addition, Kay *et al.* point out it is likely that bathing water standards in European Union member states will be made more stringent in future years so that efforts to "clean up" UK coastal waters face what in past political parlance has been called a "double whammy".

Studies carried out at a popular tourist beach in south Wales by Cliff Nelson and his colleagues (Chapter 4) show that swimming in seawater contaminated by sewage increased the risk of contracting a gastrointestinal illness by 21 times. At the same time, questionnaire surveys undertaken by Nelson *et al.* revealed that the public perception of both beach and water contamination was high. More than two-thirds of visitors were aware of the water pollution problem, and poor water quality was given as the reason by more than half the visitors who did not go swimming.

In studying the best way to achieve desired water quality standards and objectives, use is often made of modelling studies because they are generally cheaper and more flexible than other alternatives, such as field investigations. However, the results of modelling studies also require careful evaluation, and models need to be rigorously validated as part of this process. In the case of models that have been developed to study the leaching of pesticides from cracking clay soils, Adrian Armstrong and his colleagues (Chapter 5) suggest that the process of validation, which involves acquiring confidence that the model adequately represents the real-world situation, will be different depending on whether the model is to be used to advance scientific understanding of pesticide behaviour in the environment or in a regulatory way to assess the environmental safety of pesticide use. In the former situation, Armstrong *et al.* specify that large amounts of data will be required to describe the system being modelled and for evaluating model outputs. However, where such information is difficult or expensive to obtain, as is the case with pesticide data, a multistage cumulative validation procedure is advocated for leaching models whereby predictions of site hydrology and solute movement are tested first and the validation confidence carried forward to a later stage when the pesticide components of the model are validated. Validation of a pesticide leaching model for use as a regulatory tool requires testing of its ability to make valid generalised predictions for a range of different sites and pesticides. Armstrong *et al.* suggest this can be achieved by repeated application of the model using data from different locations.

In order to simulate successfully the impacts of environmental change on water quality, models have to be capable of taking into account the complex and dynamic effects of physical, chemical and biological processes. Paul Whitehead and his colleagues (Chapter 6) provide an example of this approach involving the Integrated Nitrogen Catchment (INCA) model which assesses multiple sources of nitrogen in river basins based on the reaction kinetics of the principal processes of the nitrogen cycle. Whitehead *et al.* illustrate how INCA has been used to investigate the effects of fertilising upland areas and increases in atmospheric N deposition for the River Tywi in Wales. Results suggest that the nitrogen levels in the Tywi are relatively insensitive to substantial increases in atmospheric deposition but are strongly affected by fertiliser application to grassland.

Application of complex physically based models does not provide a simple means of simulating water quality on a national scale. Paul Quinn and his colleagues (Chapter 7) show how Minimum Requirement Models (MIRs), which are simple, functional low-parameter models built specifically to mimic the key outputs from more complex models, but using the minimum amount of available data, can be used to predict the possible range of N losses in UK agricultural catchments. Quinn *et al.* argue that these models have a number of advantages including easy use across a range of catchment scales, evaluation of model uncertainty facilitated by a simple structure, and ready integration with a national Geographic Information System (GIS) database of agricultural statistics.

LINKING SCIENCE AND POLICY

Advances in the scientific understanding of water quality are used to inform the development of policies and practices for managing water resources. Pauline Kneale and Adrian McDonald (Chapter 8) suggest that for upland areas in northern England water quality problems associated with manganese enrichment, algal blooms, coloured runoff and *Cryptosporidium*, together with increased demand and shorter reservoir storage times, increasingly threaten potable supplies. However, they argue that coupling of existing process understanding with easily updated GIS-based monitoring systems provides a powerful water management tool through which hazards can be identified and risks assessed.

In some countries it has taken the occurrence of a major water quality crisis to stimulate not only the formulation of policy but also the basic scientific research on which management strategies can be based. David Ingle Smith (Chapter 9) describes how major algal bloom affected the Darling River, Australia, for a distance of 1000 km during the summer of 1991/92. This rendered supplies of stock water unfit for consumption and cost many millions of Australian dollars in provision of emergency filtration equipment and lost revenue from tourism. It also led to reforms in the economic and institutional arrangements relating to water allocation, pricing and inter-state transfers, which stress the importance of maintaining environmentally appropriate flow levels. Smith also points out that the serious algal contamination in the early 1990s highlighted a general paucity of data on all aspects of water quality and stimulated an upsurge of research to provide a sound science base for algal management.

Integrating scientific understanding into strategies to protect the quality of water resources may not always be easy. Chris Soulsby and his colleagues (Chapter 10) review groundwater and pollution risk in the UK and conclude that while it is possible at a strategic level to define the general vulnerability of an aquifer based on soil and rock properties, the development of protection policies related to such classification is constrained by political and economic considerations. Similarly, at a local level, Soulsby *et al.* suggest that while there are many scientific approaches for delineating protection zones around individual groundwater sources, including sophisticated models of groundwater movement, their implementation requires subjective judgement to incorporate an adequate margin of safety at an appropriate economic cost.

The relationship between the base of scientific understanding and the development of water quality policy is often a complex and dynamic one. In the case of water pollution associated with urban environments, for example, Bryan Ellis and Bob Crabtree (Chapter 11) demonstrate that policy and practice in approaches to urban surface stormwater management in England and Wales do not reflect advances in the scientific understanding of a source control approach to managing pollution through detention storage, wetland systems, infiltration trenches, porous paving and grass swales. In contrast, they argue that present urban wastewater management policy reflects sound science-based approaches to pollution control, but new policies, which will be needed to ensure future environmental improvement, will require further advances in the understanding of wastewater system performance and associated environmental impacts.

There are several factors, in addition to scientific understanding, which influence the management and remediation of water contamination problems, including economic constraints. Colin Green (Chapter 12) points out that the economics of water pollution abatement are complex and not always readily treated in terms of neoclassical economic analysis in which "use", "functional" and "non-use" values are attributed. Green uses a case study of discharges from gold mines in the East Rand Basin, South Africa, to illustrate problems of deciding on the best economic option for abating pollution of a Ramsar designated wetland. It was concluded that neither closing the mines nor implementing expensive desalination treatment of the discharge could be economically justified. Rather, continuing with the aeration, liming and clarification of the effluent was considered the best approach.

While regulations and legislation, based on sound scientific principles, may be framed to control water contamination, there may not be a willingness among all potential polluters to take appropriate remedial action. Ian Foster and his colleagues (Chapter 13) show from a survey conducted in south Warwickshire that farmers have a general lack of interest in problems associated with inorganic farm waste and in many cases are ignorant of, or blatantly disregard, existing legislation and codes of good agricultural practice regarding its disposal. It seems that economic, rather than environmental, considerations dominate farmer behaviour, and that there is a reluctance amongst the farming community to seek advice and guidance from the regulatory authorities regarding waste disposal, not least because of a fear that too close an inspection of existing practices will incur prosecution.

Where regulations are breached and water pollution occurs, regulatory authorities have the option of imposing financial sanctions on polluters to effect clean-up and to secure future compliance with water quality standards. While this approach has generally been successful in the western world, Stuart Lane and his colleagues (Chapter 14) argue that in countries such as India, there may have been a tendency to rely too heavily upon a sanctions-based approach to water pollution control. Despite the existence of apparently powerful laws of protection, Lane *et al.* point out that they have not been properly enforced, and continued water pollution is a feature of the country. This reflects factors such as the history of pollution control, lack of awareness of contamination problems, economic imperatives and the role of the legal process. They argue that a compliance-based approach, in which there is more negotiation between polluters and regulatory authorities to improve water quality, may be a more effective approach to pollution control, although it is also recognised that the success of this alternative depends on the willingness of different institutions within India to adopt a more co-operative stance.

In the final chapter of this volume (Chapter 15), Tim Burt suggests that there is a need for an integrated approach to land-use planning in rural drainage basins within the UK, involving consensus rather than conflict between the individual and organisational stakeholders concerned. Burt argues that such an approach would facilitate the strengthening of environmental considerations in the decision-making process and, in turn, would promulgate the adoption of more sustainable solutions to water quality management. While a limited number of examples of integrated catchment management can be cited from the UK, Burt suggests that a more extensive adoption of this approach, along the lines of the Landcare movement in Australia, offers a promising way forward.

REFERENCES

Meybeck, M., Chapman, D. and Helmer, R. 1989. *Global Freshwater Quality. A First Assessment.* Basil Blackwell, Oxford.

Peters, N.E., Bricker, O.P and Kennedy, M.M. 1997. *Water Quality Trends and Geochemical Mass Balance.* John Wiley, Chichester.

Peters, N.E., Bonell, M., Hazen, T., Foster, S., Meybeck, M., Rast, W., Schneider, G. Tsirkunov, V. and Williams, J. 1998. Water quality degradation and freshwater availability – need for a global initiative. In *Water: A Looming Crisis? World Water Resources at the Beginning of the 21st Century,* Proceedings of the International Conference, 3–6 June 1998, Paris, France, UNESCO, 195–201.

Section One

GLOBAL PERSPECTIVES

1
Water Quality: an Emerging Global Crisis

Edwin D. Ongley

INTRODUCTION

Water exercises a pervasive influence over all aspects of human existence, yet is the most ignored of all commodities in the management of national economies. Water is almost never considered a "sector" in the economy, is infrequently managed holistically by governments, and rarely is assigned a relevant economic value in sectoral planning. Historically, water quantity has been managed through supply-side policies that have emphasised development of new water resources rather than managing demand. Lakes and rivers the world over are used as free waste-disposal facilities. Groundwater, which is out of sight and usually out of mind by government agencies, is being over-pumped and contaminated by urban, agricultural and industrial wastes almost everywhere.

Increasingly, experts are referring to the global water situation as a "crisis" which will have its major impact early in the twenty-first century. While the water quantity crisis is well known in areas such as the Middle East, there are many professionals who believe that *freshwater quality* will become the principal limiting factor for sustainable development for many countries early in the next century. Indeed, in countries such as China officials are, for the first time, publicly stating that water quality is now limiting economic development.

On a global basis, the water quality "crisis" has the following global dimensions (Ongley, 1996):

Water Quality: Processes and Policy. Edited by Stephen T. Trudgill, Des E. Walling and Bruce W. Webb.
© 1999 John Wiley & Sons Ltd.

- five million people die annually from water-borne diseases
- ecosystem dysfunction and loss of biodiversity
- contamination of freshwater and marine ecosystems from land-based activities
- contamination of groundwater resources
- global contamination by persistent organic pollutants.

Economically and politically, these can lead to (Ongley, 1996):

- decline in sustainable food resources (e.g. freshwater and coastal fisheries) from pollution;
- cumulative effect of many countries making poor water resource management decisions because of inadequate water quality data;
- many countries can no longer manage pollution by dilution, leading to higher levels of aquatic pollution;
- escalating cost of remediation and loss of "creditworthiness";
- international instability;
- environmental refugees.

THE INTERNATIONAL INSTITUTIONAL FRAMEWORK

Recognition of this situation is certainly not new. As early as 1972 at the first United Nations Conference on the Environment held in Stockholm, the status of the world's water resources was discussed and led to the formation of the United Nations Global Environment Monitoring System (GEMS) with a specific water quality component "GEMS/Water", with its principal collaborating centre sponsored by the Canadian Government at the Canada Centre for Inland Waters. In 1977 a major international conference on water resources was held in Mar del Plata, Argentina, and produced the first major road-map of steps and policies that were needed to address the rapidly deteriorating situation at global and national levels. Unfortunately, few of the recommendations were taken to heart by governments, and especially by developing countries where the most serious water problems are now emerging. A major global initiative, the International Decade on Drinking Water Supply and Sanitation (1980–90) saw significant progress in providing clean drinking water to urban areas in the developing world. However, population increase and massive rural to urban migration has led to a situation where it has not been possible to achieve significant relative improvement over the decadal period.

In the decade of the 1990s, several key events have led to the current situation of 1997. The Swedish Government together with the Stockholm Water Company began an annual Stockholm Water Symposium together with the Stockholm Water Prize ($150 000) to provide an annual forum for discussion of major water issues. The seventh Symposium was held in 1997 on the theme "With Rivers to the Sea". The International Conference on Water and the Environment was held in Dublin in 1992 as a planning mechanism for the water sector for the ensuing Earth Summit (United Nations Conference on Environment and Development – UNCED) held in Rio de Janeiro later the same year. Chapter 18 (the Water Chapter) of Agenda 21, which was adopted in Rio, provided a comprehensive statement and recommendations on management of freshwater.

In 1994 the Dutch Government sponsored the Ministerial Conference on Drinking Water and Environmental Sanitation in Noordwijk, The Netherlands, which reaffirmed the need to implement Chapter 18 of Agenda 21 and outlined an Action Plan. The importance of controlling land-based sources of pollution to marine environments led to the 1995 approval by governments of the United Nations' "Global Programme of Action for Protection of the Marine Environment from Land Based Activities" (known as the GPA). In 1996 the World Water Council was formed to provide a forum for debate by governmental and non-governmental organisations (NGOs) on water. Also in 1996 the Global Water Partnership (GWP) was established under the leadership of the World Bank and the Swedish Government to provide a co-ordinating mechanism for donors and recipients for delivery of technical and financial assistance in the water sector. At the Governing Council of the United Nations Environment Programme (UNEP) in January 1997, several countries introduced a successful resolution that elevated freshwater into a much greater profile in the UNEP agenda.

Although the water crisis was flagged as early as 1977 at the Mar del Plata conference, water has had difficulty surfacing at the political level. In part, this has reflected the national view that water management was a national sovereignty issue. A second reason lies in the fragmented and sectoral institutional responsibility for water management that continues in most countries and, by extension, the lack of a unifying national water management policy. Without a single and visible authority responsible for water, the issue has largely been ignored at the national policy level. It is not surprising therefore that the water crisis was largely in the domain of water professionals and not on the table of international politics.

Therefore, and despite so much international activity by professional groups, United Nations specialised agencies, donor countries and NGOs, most national governments were neither significantly involved in the water issue, nor particularly aware of the looming freshwater crisis. It was not until the Interim Session of the Commission on Sustainable Development (CSD) in April 1997 and the following Special Session of the United Nations General Assembly that there was a major turning point in the politics of water. At these meetings the world's governments agreed that freshwater was a priority global issue and identified it as a major topic of intergovernmental discussions at the 1998 meetings of the CSD and the United Nations General Assembly. Freshwater had finally achieved significant visibility within the global political agenda in 1997.

DEFINING THE PRESENT SITUATION AND ANTICIPATING THE FUTURE

There is a rapidly growing dichotomy between the water quality problems of the developed versus the developing world. This dichotomy leads to serious problems of technology transfer and even of the data paradigm that usefully applies to developing countries. Generally, the developed world has moved beyond problems of simple eutrophication and faecal pollution, to problems of aquatic ecosystem protection, to risk-based toxicity and effects assessment at the organism, cell and DNA levels, to endocrine disruption and inter-generational effects of toxic contamination. In developed countries, public opinion is well developed and political systems respond to this pressure. Most importantly, aquatic and environmental quality is considered a public good and worthy in its own right. These values are accompanied by sufficient financial and human resources to make most things possible.

In contrast, developing countries are faced with the simultaneous impact of old issues such as faecal contamination and impacts on public health, with more recent issues of eutrophication and acidification, and with new issues of toxic contamination. All of these have impacts both on surface water and groundwater resources. At the same time, these countries rarely have the human resources and experience or the financial resources to effectively deal with this range of complex issues. This is complicated by the lack of a well informed public to become involved in a political process that will influence the national environmental agenda. National priorities are invariably focused upon economic development, usually at the expense of the environment. Recently, however, there is evidence that some rapidly developing countries have recognised that water quality is now placing serious restrictions on economic development and they are preparing to invest heavily in remediation. Apart from the financial implications, their difficulty lies in their inexperience in dealing with complex aquatic issues, the lack of trained personnel in such issues, and an inefficient and complex institutional situation.

The findings of the Comprehensive Assessment of Freshwater Resources of the World (United Nations, 1997) paint a compelling and dismal picture of the future of global freshwater resources. That study uses two important concepts, "water stress" and "freshwater vulnerability".

The term *water stress* is defined as the ratio of water withdrawal to water availability on an annual basis. The United Nations defined four levels of water stress.

1. *Low water stress*: less than 10% of available freshwater is used.
2. *Moderate water stress*: water use is in the range of 10–20% of available water; availability is becoming a limiting factor.
3. *Medium-high water stress*: water use is in the range of 20–40% of available water; management of both supply and demand will be required to ensure that the uses remain sustainable.
4. *High water stress*: more than 40% of available water is used; this indicates serious scarcity and is often accompanied by unsustainable use of groundwater resources.

However, the ability of nations to cope with water stress depends greatly on income and stage of development. Therefore, the term *freshwater vulnerability* describes the combination of water stress and income levels. The result is a series of categories showing the vulnerability of various countries and regions to problems caused by water scarcities. Most important is the United Nations' observation that "over one half the world [population] falls in the low income category, and more than one third of these people are in countries that already face medium-high to high water stress."

Whereas the world now utilises some 54% of the available resource, economic predictions indicate that by 2025 fully 100% of the available resource may be required to meet the range of demands for water by agriculture, industry and urban needs (United Nations, 1997). This projection does not factor in the need for water to maintain healthy ecosystems and suitable levels of aquatic biodiversity. Of course, full utilisation of the world's freshwater resources is not even possible as it assumes that water can be transported between regions and continents, and that national water use is managed comprehensively and effectively.

The future for water quality is even more problematic than for water scarcity. The United Nations Industrial Development Organization (UNIDO, 1996) has estimated that biological pollution loading, suspended solids and total dissolved solids loadings could increase by the year 2025 by up to 18.5, 16.8 and 17.2 times 1990 levels, respectively, for the dynamic Asian economies. For the rest of Asia the increases are estimated to be five to ten times 1990 loadings. For those familiar with Asian situations the prospect of such a major increase is almost unthinkable. The problem with water quality is that, unlike water quantity for which scarcity at the national level can be captured in simple and easy-to-comprehend statistical and economic terms, water quality has mainly local impacts and is difficult to summarise at national levels.

The linkage between water scarcity and water quality is extremely close. Because of the unavailability of new and economically viable freshwater sources in most parts of the world, global views of alleviating water scarcity depend mainly upon the following technical measures:

1. greatly improved efficiency of water use at the basin level, including improvement in efficiency at the sectoral level (agriculture, municipal and industrial uses); and
2. reduction in water pollution insofar as pollution adds to water scarcity through loss of beneficial use.

Although not well developed, a third issue that is now being promoted in some quarters is the need to examine world food markets relative to national food security policies with the objective of increasing the trade in "virtual" water (that is, the water represented in grain imports/exports and other food products) from water-rich regions to water-poor regions.

Water Quality and Food Security

Even now the role of agriculture in water quality is substantial. Although few countries are able to quantify the role of agriculture in national water pollution, the United States is an example of a developed country where agriculture is the major polluting source for surface and groundwater for a broad range of substances (Table 1.1). This is typical of countries where point sources have been broadly regulated. There has been widespread concern in Europe for several decades over increases in nitrogen, phosphorus and pesticide residues in surface and groundwater.

In developing countries there is no doubt that point sources are the major impacting sources; however, the role of agriculture and other types of non-point sources is not

Table 1.1 *Leading sources of water quality impairment in the United States. From US-EPA (1994)*

Rank	Rivers	Lakes	Estuaries
1	**Agriculture**	**Agriculture**	Municipal point sources
2	Municipal point sources	Urban runoff/storm sewers	Urban runoff/storm sewers
3	Urban runoff/storm	Hydrologic/habitat modification	**Agriculture**
4	Resource extraction	Municipal point sources	Industrial point sources
5	Industrial point sources	On-site wastewater	Resource extraction

known and may be substantial. The absence of reliable data makes the assessment of agriculture relative to point sources difficult or impossible in such countries. Moreover, the presence of large shallow lakes with large internal loadings of phosphorus, especially in Asian countries, is a major complicating and often overlooked factor when remediation projects are planned.

The debate over the freshwater scarcity issue is greatly complicated by the global debate over the future of food security. Water used in agriculture amounts to some 70% of total water withdrawals; agriculture is responsible for 93% of total water consumed by all economic sectors (United Nations, 1997). Nevertheless, the Food and Agriculture Organization of the United Nations (FAO) (Alexandratos, 1995) has provided a somewhat reassuring picture of the world's ability to produce enough food to meet the demands of a growing global population. Others such as Brown (1996) take a much more pessimistic view, reflecting observations that:

- the world population is increasing at some 90 million annually;
- demand for grain is increasing faster than population growth due to dietary changes by Asian populations;
- water supply is fixed, yet increasing amounts will be diverted from agriculture to industrial and municipal uses;
- there is no large amount of additional land available either for irrigated or rainfed agriculture;
- benefits of the "green" revolution have levelled out;
- grain production per capita is now declining;
- the major world fisheries are either in absolute decline or are being fished unsustainably;
- increasing freshwater pollution may limit expansion of freshwater aquaculture.

The food security issue has the following water quality implications (Ongley and Kandiah, 1997).

1. Intensification of production both of rainfed and irrigation agriculture and of aquaculture will lead to increasing levels of fertiliser and pesticide runoff.
2. Further expansion of rainfed agriculture into marginal lands that are highly susceptible to erosion will increase sediment runoff and freshwater turbidity and siltation.

3. The need to rehabilitate salinised irrigation lands and to more effectively utilise salinised land will add to salinity loadings to aquatic systems.
4. Intensification of livestock raising, especially in Asia, to meet increasing demand for protein will result in increasing loadings of faecal matter, organic and inorganic wastes.
5. Further expansion of the agro-food processing industry will increase loadings of biological and chemical oxygen demand.

Ecosystem Health and Water Quality

Aquatic ecosystem health, while simple to understand in the abstract, is difficult to assess or predict in meaningful terms and is closely linked to water quality. The science behind ecosystem health is difficult and not fully developed, and the political accept-ance of a meaningful implementation of the concept is difficult. Indeed, whereas aquatic biodiversity is politically acceptable as a public good in developed countries, such a concept is not generally implementable in most developing countries where basic public health needs and economic development are the priorities and where environmental needs are low on the list of priorities. Nevertheless, it is recognised by some developing countries that degraded (aquatic) ecosystem health, however that may be defined, is causing systemic failure in economic planning and development. How to deal with restoration of aquatic systems remains, however, a scientific and policy dilemma for such countries.

THE POLICY DILEMMA

The international debate over water in general and for water quality in particular has tended to be polarised into: (a) the developing nations' view that they are discriminated against by developed countries by their failure to provide adequate financial aid and technological assistance and capacity building; and (b) the developed nations' view that water policy and water conditions are primarily the result of policy and institutional failure within developing countries and that the solution is not to provide more aid, but rather a more efficient and effective use of existing internal resources, foreign direct investment, good governance, a well managed economy and appropriate environmental regulations. Of course, both views are true and there is some hope that this polarisation is now breaking down as both developing and developed countries recognise that failure to deal with the situation has profound domestic as well as global implications.

Although there are many factors that contribute to the global water quality crisis including factors such as lack of funds, lack of access to appropriate technologies, and inadequate expertise at the national and local levels, the fundamental root cause is institutional and policy failure at national levels. Only when this is recognised and accepted by national governments will there be the opportunity to make significant change through the processes of financial and technical aid and capacity building. As an example of the factors that inhibit effective national water quality management and which are intrinsic to national policy development in this sector, the following were summarised by Ongley (1997b).

1. *Policy and institutional failure.* This is the root cause of much of the crisis in developing countries. It transcends many of the following issues, but it must be highlighted because without policy and institutional reform, the water crisis will never be resolved. This element also encompasses essential legal changes that empower institutional activities.

2. *Data crisis.* Water quality and quantity networks are failing to provide the kind of information governments need to develop, implement and monitor water policies and programmes. For water quality, these are highly inefficient and ineffective, often are duplicated in two or more government agencies, are expensive to operate, and fail to provide the kind of information necessary to develop control options, or for investment into remediation programmes. This crisis extends from data collection, to data management and deployment for decision purposes both for point and non-point source management.

3. *Planning and investment: the relative roles of government and the private sector.* The relative roles, together with pros and cons, need further clarification. Cons include, for example, private sector mechanisms that fail to consider traditional societal values and which often ignore equity issues.

4. *Capacity building.* There is a large amount of ineffective and wasteful capacity building. This is the fault both of donors and recipients. Rules of engagement need to be established with objective measures developed for scoping needs, for defining the nature of capacity building that is effective and sustainable, and for means of measuring success. Capacity building also needs to focus on core competencies that do not exist at the national level (especially considering point number 7) and that are essential for efficient and effective planning and decision making. Another aspect is the failure of many internationally funded programmes to build appropriate capacity at local levels that can, in turn, be used in similar projects in the same country.

5. *Economic evaluation.* A root cause for failure to take action by national governments is the lack of any compelling information on the economic cost and lost opportunity caused by policy and institutional failures in the water quality sector, and by loss of beneficial use of polluted water.

6. *Technology transfer.* Although World Wide Web access to databases (e.g. clearinghouses) is growing dramatically, the problem is now one of how to make the information usable. Accessing knowledge and its use in decision making remains extremely difficult for developing countries. Specific attention needs to be paid to new information technologies that permit user-friendly use of knowledge bases (as well as data) for decision making, for planning, development and issue-specific management.

7. *Institutionalised competition between decision makers and local competencies in science and technology.* In many countries, incompetent national and local authorities unfairly compete directly with domestic scientific establishments for internationally funded contracts and related opportunities. The result is that inadequate and sometimes bad science is used by decision makers who fail to recognise and use excellent domestic expertise. A secondary outcome is that capacity usually fails to be increased significantly at the domestic level because the bureaucratic institutions are not able to further develop or deploy the

knowledge base once the project is over. In some countries this is the reason why a viable domestic capacity in environmental consulting has not developed.

OTHER FAILURES

The Data Paradigm

Ongley (1994) defined the water quality data paradigm as:

> The data paradigm, especially in developed countries, is data intensive, chemistry focused, science driven, high tech, well funded, and rests on the scientific and legal premise that with enough data all water quality issues are capable of an unambiguous resolution. In the United States the paradigm reflects a highly litigious society in which "more and better" data is the currency of the legal process. Protocols of US agencies reflect this paradigm... Even in the United States there is now recognition that this paradigm has led to a "data rich, information poor" situation which public agencies are now vigorously attempting to change.

This data paradigm has proven not to be especially useful even in developed countries where the cost and validity of this approach are being challenged on the grounds that conventional monitoring data: are difficult to interpret in terms of ecosystem or public health effects (presumably what monitoring programmes are intended to do!); fail to capture presence/absence of many toxic chemicals; provide little information on toxicity; provide information that is neither sufficiently linked to real management issues nor provides sufficient detail to meet scientific data needs; are data-driven rather than service-driven (that is, data should fulfil a service function); and are almost never subjected to cost–benefit analysis. The implications for developing countries of adopting this "western" paradigm are particularly severe and very costly, as noted below.

The root cause for this uncritical acceptance of the data paradigm is institutional inattention and management acceptance of conventional operations and practices of data programmes. This does not deny the fact that many countries, especially developed countries, spend much time and money to ensure data quality. The fact remains, however, that programme managers rarely challenge the fundamental premises of the data paradigm and, consequently, remain hostage to a way of doing business that is excessively costly and would never meet the criteria of cost versus benefit. The implications, especially in developing countries, can be very substantial including large costs for programmes that produce little useful information, data that are incomplete and unreliable, and the costs of making policy and investment decisions on incorrect information or having to mount new programmes to remedy data short-comings.

A specific problem of the data paradigm is its central dependency on the concept of "accuracy". The primary objective of monitoring programmes is to produce "accurate" analytical values. While accuracy is of paramount importance in data programmes, the slavish adherence to the western data paradigm denies the reality that natural aquatic systems are so variable, and monitoring programmes so unable to capture this variability, that "accurate" laboratory values often have little meaning in the real world. As an example, Ongley (1993) found that random numbers were as useful as actual chemical values for the determination of chemical loadings (used for many water quality management situations), yet agencies continue to fill their databanks with expensive ND (not detectable) values. It is urgent that governments begin to manage aquatic systems by a judgemental process that accepts that data programmes will never be able to capture all the variability in the natural system, and that data programmes will never provide a deterministic (exact) solution. The alternative is informed judgement based on sound scientific understanding and a data programme that can be used to determine water quality status and to monitor the direction of change of the system in response to managerial actions. Monitoring for regulatory compliance has different objectives and practices and is not discussed here.

The Data Crisis

The data crisis has many dimensions, and follows from the data paradigm. Ongley (1994) described the situation as follows.

> ...a common observation amongst water quality professionals is that many water quality programmes, especially in the developing world, collect the wrong parameters, from the wrong places, using the wrong substrates and at inappropriate sampling frequencies, and produce data that are often quite unreliable; the data are not assessed or evaluated, and are not sufficiently connected to realistic and meaningful programme, legal or management objectives. This is not the fault of the developing world; more often it results from inappropriate technology transfer and an assumption by recipients and donors that the data paradigm developed by the developed world is appropriate in the developing world.

Many developing countries are unable for institutional, financial and technical reasons to mount stable, reliable monitoring and assessment programmes. Water quality monitoring is often fragmented amongst several government agencies such as ministries of health, industry, transportation, energy and agriculture. In too many countries there has been a virtual collapse of systematic data programmes for water quantity and quality. In most developing countries and some developed countries there are no national data standards to ensure data quality and it is assumed (usually incorrectly) that legislated laboratory quality control as part of the analytical process will suffice. Data unreliability, including intentionally fraudulent data, is all too common. There is a profound lack of data, especially on man-made organic and inorganic compounds of industrial and agricultural origin in most countries outside the developed world. In many countries data holdings remain on paper records only and are unavailable in electronic database format.

For regional and global assessment purposes, the data crisis is serious. It has been impossible to carry out comprehensive assessment of, for example, nutrient or contaminant status in large parts of the world. The linkage with global issues such as biodiversity or source identification of toxic chemicals that are transported long distances by atmospheric processes, cannot be established. Consequently, effective solutions are difficult to derive. Loadings of pollutants to oceans and coastal and inland seas from the world's rivers are poorly known. For domestic purposes the data crisis is verging on catastrophic for many countries. National governments and river basin agencies do not have the data required to develop effective policies for water resource planning, for pollution abatement and remediation, for cost-effective source control, or for determination and application of water quality standards for maintenance of ecosystems and biodiversity. Sadly, many donors and international financial institutions fail to recognise the inadequacies of national and regional monitoring programmes.

The impact of the data crisis on agriculture and on national pollution abatement planning and investment programmes can be summarised as follows (Ongley and Kandiah, 1997).

1. Serious widespread degradation of water resources that can be primarily attributed to agriculture, as in nitrate pollution of groundwater, may go unrecognised.
2. Spatially limited downstream pollution from small-scale agricultural practices such as animal-raising are not recognised yet may have severe public health consequences.
3. The presumption of widespread pollution from agricultural practices may not be warranted relative to other contributors of nutrients, sediment and toxic substances.
4. Prediction of downstream pollution for new agricultural development schemes (e.g. irrigation expansion) is not possible.
5. Investment decision making for point or non-point pollution control is unable to assess the cost–benefit of different pollution abatement options.

At the basin level, the problem typically is (Ongley, 1996):

1. Environmental status: features of a highly eutrophic or contaminated lake or river may include high turbidity, "smelly" sediments, low oxygen, excessive growth of aquatic weed and ecosystem dysfunction.
2. The database and institutional capability are very frequently found to involve: no point or non-point source controls; little relevant data; poor laboratories; inadequate science/knowledge of issue; little money.
3. The usual questions in such situations are: What are the relative impacts of point and non-point sources, including agriculture, for nutrients, sediment, pathogens, salinisation and contaminants?
4. Data programmes can diagnose the nature of simple problems, but are usually unsuitable for designing remediation options, for developing policy alternatives, or for directing investment programmes.

SOLUTIONS

Getting Their Attention (Ongley, 1997c)

One of the major problems in moving the political agenda in regards to water quality has been the failure to create a political awareness of water quality. While the international community has recognised that freshwater is a central issue in social and economic development, the freshwater "crisis" is still largely thought by governments to be a quantity issue. This reflects the facts that: (1) water quantity and water supply have received much political and donor attention over the past years; (2) data on water quantity are more easily accessible than for water quality; and (3) water scarcity can be conceptualised at a national level within the framework of national water/allocation policies. In contrast, and although water quality is now regarded by many countries as a serious limiting factor to economic and social development and to ecosystem maintenance, the effects of impaired water quality are felt at a local level, and there is no simple way to characterise the impacts of poor water quality at an aggregated national level.

The consequence is that national governments tend to be unaware of the aggregate impacts and associated economic implications of water quality deterioration at a national level. Furthermore, without the knowledge of economic losses associated with degraded water quality in the various economic sectors (e.g. agriculture, industry, public health), national or regional governments have no basis to develop a cost-effective national or basin-wide remediation and investment strategy for water quality.

Traditionally, global water quality assessments attempt to identify the types and magnitude of impairment of water quality, such as eutrophication or water-borne diseases, by the analysis of water quality data sets. This leads to a comparative list of the characteristics of water quality pollution. While this type of approach provides a useful comparison of the nature and extent of different types of water quality, it is limited by data gaps that characterise many countries, and by the descriptive nature of the analytical process which, for the most part, cannot illuminate the root causes nor the economic implications of the observed water pollution impacts. There is no doubt that this type of descriptive approach is valuable for comparative *global* summaries of the nature of water quality impairment; however, judging by the lack of political awareness of water quality as a significant national (as opposed to local) issue, the descriptive approach has failed to provide the compelling evidence for water quality policy change at the *national* level.

This failure is probably linked to the fact that national leaders fail to understand the magnitude of the economic cost of water pollution to their national economies. It has been demonstrated that it is possible to develop an economic profile of the cost of water pollution to the national economy. For China, the only country where (to this writer's knowledge) this has been attempted, the aggregate cost of water pollution to the national economy was estimated to be 0.5% of gross domestic product or, in dollar terms, more than the value of all exports from China in the 1990 base year (Smil, 1996). As one means of focusing political attention on the very large cost of water pollution to national economies, similar analysis for other countries would provide a common and comparative analysis of gross costs, as well as providing a rational basis for national governments to focus water policies on those issues that have maximum concern at national and basin levels.

Modernisation

Modernisation of water quality programmes is an urgent priority for national governments. The principal components of modernisation include technical, institutional and legal aspects, and capacity building, which were summarised by Ongley (1997a):

Legal and institutional considerations
- role of government in water data programmes
- commitment by management to change
- changing legal and regulatory standards for greater efficiency
- building a constituency

Technical
- laboratory programmes
- multiple techniques within monitoring programmes
- new screening and diagnostic "tools"
- data quality objectives
- optimisation of the national network
- information systems
- quality control/quality assurance/accreditation/GLP (Good Laboratory Practices)
- reporting

Capacity building
- training
- institutional development: (a) client-oriented programmes; (b) revenue generation.

Capacity Building (Ongley, 1997d)

As noted above, the root cause for water quality impairment in most developing countries lies not so much in the lack of data, but rather in the failure of institutions and policies that deal with water quality management, regulation and protection. While one can always argue for scientific programmes and for new knowledge, the issue for most countries is not the need for new knowledge (indeed, little new knowledge is needed to manage the majority of water quality issues), but need for building of personnel and institutional capacity in the methods of developing and applying water quality programme elements to real water management issues. The need is not for new science or new methodologies; the need is for transfer of existing knowledge and modern methodologies.

There are many fundamental problems with capacity building programmes in the water quality sector. Too often these are a collection of short courses which donors are able to provide, and which fail to take into account the question of sustainability once the donor leaves. Short courses tend only to be useful for (a) very specific technical issues that are targeted to specific technical individuals, and (b) familiarisation courses for senior managers on general topics of water management etc., and with the objective of raising the level of understanding of water quality management issues. The worst kind of capacity building is when water agencies in developing countries take advantage of any course(s) that happen to be available by donors at low or no cost; while cost is certainly an issue, this often produces uncritical and unfocused training that does not serve the interests of the programme.

Capacity building must also be accompanied by a programme of recruitment and promotion of well trained junior professionals. Commonly, in developing countries, water organisations tend to be managed by older individuals who are appointed for political reasons, and who may not have modern skills; moreover, such organisations often fail to provide opportunity for promotion of younger and better trained persons to positions of responsibility. Institutional stability is also a factor insofar as well trained individuals are often lost to the organisation because of lack of internal opportunity and better salaries in the private sector. Capacity building is, therefore, a complex task requiring evaluation of institutional human resource management in all its aspects.

Two other aspects of capacity building also create problems that need to be managed. One is associated with "tied aid" insofar as the target country must use consultants from the donor country. Frequently, this leads to inappropriate technology transfer, often by consultants who are insensitive to, and not knowledgeable of, the local circumstances and conditions. The second aspect is infrastructure development, especially that associated with apparatus. Frequently, capacity building focuses on supplying new apparatus which, in too many instances, is unused because of lack of spare parts and unavailability of supplies. This problem can be remedied, but requires a "sustainability plan" on the part of the recipient and donor.

A central area of capacity building in the water quality sector that is not handled well by donors or recipients, is the fact that water quality management is a complex issue that involves a wide range of needs. As an example, if country "A" wants to proceed with a remediation plan for a large river–lake system, there is a range of institutional, scientific and technical, and programme requirements that need to be considered. Most developing countries are not well equipped to handle such complexity within a holistic context. Country "A" may well have excellent hydrologists and sedimentologists, but does not have skills in areas such as aquatic ecology, risk assessment and comprehensive basin planning. Because basin remediation requires a set of trade-offs amongst the various aquatic components and amongst users, water quality professionals in, say, laboratory operations need also to understand how the laboratory programme fits into the larger set of issues and clients. Similarly, basin planners need to fully understand the limitations of science in developing and predicting systemic shifts in water quality and ecological systems so that the basin plan is reasonable, defensible and capable of achieving its objectives.

There is an urgent need for capacity building programmes that take a comprehensive and holistic view of water quality management. What is needed is a demonstration programme associated with real and typically complex water quality management issues in a selection of developing countries. A demonstration programme would provide a regional focus for trainees from adjacent countries. The objective would be to provide training through hands-on involvement in real issues in the following topical areas which are not well understood in developing countries.

1. *Developing data quality objectives.* This is a process of scoping the information needs of water quality managers. This may include a large range of potential "clients". The implication is that water quality programmes provide a service to clients, and should be service-driven rather than data-driven.

2. *Dealing with complexity of issues.* Typically in Asian countries, for example, remediation of freshwater systems involves trade-offs between competing interests such as point source versus non-point source impacts, or aquaculture as an economic necessity versus pollution impacts. A major failure of water quality programmes is the lack of recognition or rationalisation of the competing demands for water quality information that can lead to optimisation of water quality management. This tends to occur because of the lack of inclusiveness of water data programmes which, in turn, reflects institutional failure.

3. *Modernising the data collection process.* There has been a revolution in monitoring practices and technologies, especially in the use of cheap screening tools. Agencies need to learn how to design, modernise and implement data programmes so that client needs are served. Modernisation also includes the use of modern laboratory methods and of alternative biological measures that are appropriate for that country.

4. *Data mobilisation.* Discussions of data mobilisation in most agencies of developing countries tend to be limited to database development. Agencies need to learn how to mobilise data for client use and how to create data products that appeal to the public and convey cogent messages to politicians and senior bureaucrats. Data mobilisation also includes the process of systematic analysis of data holdings in order to modify and improve the data collection process.

5. *Decision support.* Beyond simple data mobilisation lies the larger issue of decision-support technologies for managing complex water quality issues. Modern decision-support technologies offer very cost-effective methods of transferring expert knowledge to the hands of users.

6. *Building a constituency.* Water quality programmes in many countries suffer from the lack of any constituency. This manifests itself in lack of support from senior officials of water agencies (who are often civil engineers or hydrologists), or support at the political level. Constituency building becomes an important element to the success and value of a water quality programme.

7. *Legal issues.* A demonstration project should include a consideration of the legal framework within which water quality programmes exist. First and foremost is the concept of national data standards which is the root cause of the data crisis in many developing countries. This may include use of a variety of accreditation and certification processes that can be presented as options to trainees. The legal standards of many countries prevent modernisation and simplification of water quality programmes; the options for remedying this situation need also to be explored with trainees.

REFERENCES

Alexandratos, N. (Ed.) 1995. *World Agriculture Towards 2010: An FAO Study.* FAO, Rome.

Brown, L.R. 1996. *Tough Choices: Facing the Challenge of Food Scarcity.* The Worldwatch Environmental Alert Series, Norton, New York.

Ongley, E.D. 1993. Pollutant loadings: Accuracy criteria for river basin management and water quality assessment. *Water Resources Development*, **9**(1), 39–50.

Ongley, E.D. 1994. Global water pollution: challenges and opportunities. *Proceedings: Integrated Measures to Overcome Barriers to Minimizing Harmful Fluxes from Land to Water.* Publication No. 3, Stockholm Water Symposium, 10–14 August 1993, Stockholm, Sweden, 23–30.

Ongley, E.D. 1996. *Control of Water Pollution from Agriculture.* FAO Irrigation and Drainage Paper No. 55, FAO, Rome.

Ongley, E.D. 1997a. Matching water quality programs to management needs in developing countries: the challenge of program modernization. *European Water Pollution Control,* 7(4), 43–48.

Ongley, E.D. 1997b. *Task Outline.* Presented to members of the United Nations Administrative Committee on Coordination, Subcommittee on Water Resources, on behalf of the United Nations GEMS/Water Programme Collaborating Centre, Burlington, Canada (unpublished).

Ongley, E.D. 1997c. *Proposal on Economic Evaluation of Water Quality.* Presented to the United Nations Administrative Committee on Coordination, Subcommittee on Water Resources, on behalf of the United Nations GEMS/Water Programme Collaborating Centre, Burlington, Canada (unpublished).

Ongley, E.D. 1997d. *Outline for Capacity Building.* Presented to the United Nations Administrative Committee on Coordination, Subcommittee on Water Resources, on behalf of the United Nations GEMS/Water Programme Collaborating Centre, Burlington, Canada (unpublished).

Ongley, E.D. and Kandiah, A. 1997. Managing water pollution from agriculture in a water scarce future. In *With Rivers to the Sea*, Proceedings of the 7th Stockholm Water Symposium. Stockholm International Water Institute (SIWI) Report No. 2, Stockholm, Sweden, 309–316.

Smil, V. 1996. *Environmental Problems in China: Estimates of Economic Costs.* East–West Center Special Reports No. 5, East–West Centre, Honolulu, Hawaii.

United Nations. 1997. *Comprehensive Assessment of the Freshwater Resources of the World.* Report of the Secretary General, Commission on Sustainable Development, 5th Session, 7–25 April 1997, Document E/CN.17/1997/9.

UNIDO. 1996. *Industry, Sustainable Development and Water Programme Formulation.* United Nations Industrial Development Organization, Document XP/GLO/96/062 (ISED/R.53).

US-EPA. 1994. *National Water Quality Inventory.* 1992 Report to Congress, EPA-841-R-94-001, Office of Water, Washington, DC.

Section Two

SCIENCE FOR POLICY

2

Riverine Mass Load Estimation: Paris Commission Policy, Practice and Implications for Load Estimation for a Major UK River System

Helen P. Jarvie, Alice J. Robson and Colin Neal

INTRODUCTION

Reliable long-term river mass load estimates are crucial for an understanding of the nature and sources of pollution to coastal waters and to meet requirements of environmental management and legislation. The Paris Commission (PARCOM) was established to administer the Paris Convention (Convention for the Prevention of Marine Pollution from Land-Based Sources, 1974), which came into force in 1978. The aim of the convention is to prevent, reduce and, where appropriate, eliminate pollution of the northeast Atlantic, including the North Sea, from land-based sources, riverine discharges and direct discharges of effluent. Individual contracting parties (countries which have signed up to the convention) are required to take part in the "comprehensive annual survey of riverine inputs" (PARCOM, 1992), which aims to monitor 90% of riverine and direct discharges of certain trace metals, nutrients, organic micropollutants and suspended sediments to the "maritime area" (the saline limit within the tidal river at periods of low river flow; Jarvie *et al.*, 1997). Individual contracting parties are

Water Quality: Processes and Policy. Edited by Stephen T. Trudgill, Des E. Walling and Bruce W. Webb.
© 1999 John Wiley & Sons Ltd.

responsible for estimating annual riverine mass load and are required to report these load data on a yearly basis to PARCOM. In 1990, new recommendations were introduced by PARCOM for sampling, chemical analysis and load estimation, in an attempt to standardise methods used by individual contracting parties. However, despite the introduction of standard methods, key discrepancies have been identified in the way loads are estimated (Jarvie *et al.*, 1997). These discrepancies relate to:

- the location of sampling with respect to tidal and saline limits in the lower river reaches;
- within the tidal reaches, whether sampling is undertaken under particular conditions of the tidal cycle;
- the periodicity of sampling and whether sampling is biased towards particular river flow conditions;
- the use of two different load estimation algorithms (PARCOM methods 1 and 2; see Table 2.1);
- the frequency of sampling.

A summary of national differences in sampling and load estimation strategies is provided in Table 2.1. The implications of these methodological differences for load estimates require prompt examination, given the need for reliable long-term load estimates to identify sources of pollution and the high political profile of such information in an international policy-making and legislative context.

In this paper, we use data from the River Trent to consider the likely impacts of variations in each of the above factors. To date, considerable attention has been placed on examining the effect of sampling frequency and load calculation method on annual load estimates. The accuracy and precision of different load estimation methodologies have been quantified, using real data (e.g. Dickinson, 1981; Ferguson, 1987; Walling and Webb, 1981, 1985, 1988; Webb *et al.*, 1997) and using synthetic simulated concentration data (e.g. Littlewood, 1995). Generally, the findings demonstrate increases in load precision with sample size (sampling frequency). It seems that the reliability of estimates is largely determined by the sampling frequency at a given site, whereas the calculation method represents a form of "fine tuning" of estimates (De Vries and Klavers, 1994).

The investigation presented here concentrates on the effects of (i) sampling location (in the non-tidal river compared with freshwater tidal and saline tidal locations), (ii) sampling under different tidal conditions and river flow conditions. There is currently a lack of information on the effects of such sampling strategies on load estimates, possibly owing to the paucity of suitable data sets. A case study is presented, which examines the effects of differences in sampling strategy on load estimates for the River Trent, a major UK river system flowing into the North Sea via the Humber Estuary. An extensive set of water quality data have been collected by the UK National Rivers Authority (now the Environment Agency) along the lower parts of the River Trent, in the non-tidal, freshwater tidal and saline tidal reaches. These relatively long-term Environment Agency data sets (spanning several years; Table 2.2) are used to calculate mean annual mass load estimates.

Table 2.1 National and regional differences in sampling strategies and load estimation methodologies for the major European rivers draining into the southern North Sea

Country	River	Catchment area (km²)	No. sites sampled	PARCOM load estimation method*	Location of sampling	Sampling frequency (samples per year)	Tidal condition
UK	Tyne	2400	3	1			
	Wear	1000	1	1			
	Tees	1500	1	1	Above tidal limit (freshwater unidirectional flow)	Approx. 12 (bias toward sampling at high flow)	NI
	Humber	24 000	8	1			
	Thames	15 000	10	1			
Belgium	Scheldt		1	2	Tidal saline	13–17	NA
The Netherlands	West Scheldt	20 331	1	1	Tidal	Trace metals, 11–14; Nutrients, 24–27; gHCH & PCBs, 7–11	Low tide
	East Scheldt		1	1	Tidal		
	Rhine/Meuse	188 630	2	1	Tidal		
Germany	Ems	8469	1	1	Tidal limit	12 (except P)	
	Weser	37 788	1	1	Tidal limit	12	Ebb tide
	Elbe	131 951	1	1	Freshwater limit	52	

*Load estimation algorithms:

PARCOM method 1: $\quad \text{Load} = K \left[\dfrac{\sum\limits_{i=1}^{n} C_i\, Q_i}{\sum\limits_{i=1}^{n} Q_i} \right] \bar{Q}$

PARCOM method 2: $\quad \text{Load} = K \left[\sum\limits_{i=1}^{n} \dfrac{C_i\, Q_i}{n} \right]$

where K is a conversion factor to take account of sampling frequency and units used for flow and concentration; C_i is the concentration measured in sample i; Q_i is the flow when sample i is collected; $\bar{Q}$ is the mean long-term flow over the period of sampling, based on the mean daily flow record; n is the number of samples.

NI, not available; NA, not available

Table 2.2 *The Environment Agency water quality resource for the lower Trent*

Site	Sampling exercise	Date range	Mean sampling interval (days)	Determinands available for load calculation
Winthorpe Bridge	I (non-tidal)	26.2.86 to 9.11.88	34	TON, SSC, NH$_3$
	II (non-tidal)	11.1.89 to 13.9.95	21	TON, SSC, NH$_3$, Fe, Pb, Cd, Zn, Cu, Ni
Dunham	I (unknown tide)	6.1.86 to 5.11.88	18	TON, SSC, NH$_3$
	II (unknown tide)	11.1.89 to 5.9.95	7	TON, SSC, NH$_3$, Fe, Pb, Cd, Zn, Cu, Ni
Gainsborough	I (unknown tide)	6.1.86 to 15.11.88	26	TON, SSC, NH^3, Fe, Pb, Cd, Zn, Cu, Ni
	II Ebb tide	24.1.89 to 5.10.95	28	TON, SSC, NH$_3$
	III Flow tide	8.12.89 to 1.8.95	71	TON, SSC, NH$_3$
	IV Slack high water	11.1.89 to 3.5.95	50	TON, SSC, NH$_3$
	V Slack low water	11.1.89 to 28.9.95	50	TON, SSC, NH$_3$
Keadby	I (unknown tide)	6.1.86 to 9.6.88	20	TON, SSC, NH$_3$, Fe, Pb, Cd, Zn, Cu, Ni
	II Ebb tide	16.1.89 to 14.9.95	20	TON, SSC, NH$_3$
	II Flow tide	10.5.89 to 2.10.95	61	TON, SSC, NH$_3$
	IV Slack high water	11.1.89 to 31.7.95	49	TON, SSC, NH$_3$
	V Slack low water	11.1.89 to 28.9.95	45	TON, SSC, NH$_3$

TON, total oxidised nitrogen; SSC, suspended sediment concentration

THE STUDY AREA: THE RIVER TRENT

The River Trent rises in north Staffordshire and flows northeast for 274 km to form the River Humber (Humber Estuary) at its confluence with the Yorkshire Ouse (Figure 2.1). The Trent drains a catchment area of 10 435 km², with a population of around six million, concentrated in Birmingham, Nottingham, Derby, Leicester, Burton-on-Trent, Stoke-on-Trent, Stafford, Cannock and Lichfield. The River Trent supplies the largest freshwater flow to the Humber Estuary, with a mean annual flow of 7526 Ml/d (1986 to 1995) recorded at the gauging station at North Muskham. This mean annual flow represents approximately 40% of runoff into the Humber Estuary. The River Trent receives runoff from a wide variety of land-use types, including urban runoff and industrial and sewage effluents from the major centres of population, abandoned

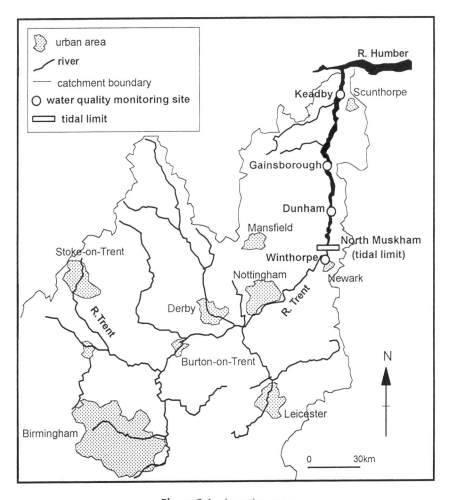

Figure 2.1   *Location map*

mines within the coalfields of Staffordshire, Derbyshire, Leicestershire, north Warwickshire and South Yorkshire, and the lead and zinc orefields of Derbyshire and iron
orefield of west Lincolnshire. The River Trent also drains important agricultural land:
intensive mixed farming takes place in the Midlands in close proximity to the urban
markets of Birmingham, Nottingham and Leicester and the fertile alluvium of the Vale
of Trent supports arable production based upon wheat, potatoes and sugar beet. The
Trent Valley also has the largest concentration of electricity generating power stations
in the country, owing to the availability of a plentiful supply of river water for cooling
purposes and the proximity of coal resources.

THE DATA RESOURCE

The reach of the River Trent considered within this study extends from just above the
limit of tidal influence at Newark on Trent, to the confluence with the Humber Estuary
(Figure 2.1). Four water quality monitoring sites were established by the National Rivers
Authority along the lower reaches of the River Trent: (1) Winthorpe Bridge (freshwater
non-tidal), (2) Dunham (freshwater tidal), (3) Gainsborough (freshwater tidal), (4)
Keadby (tidal with some saline influence). A series of separate sampling exercises was
undertaken at these sites between January 1986 and October 1995, facilitating comparison of loads at different locations along the tidal and non-tidal lower reaches. At Keadby
and Gainsborough, chemical concentration data are available under different tidal
conditions, facilitating examination of the effects of biasing sampling to certain times
within the tidal cycle (ebb tide, flow tide, slack high and low water).

   The data set includes a wide range of inorganic chemical determinants, including
trace metals, nutrients and suspended sediment. A summary of the chemical concentration data available for load calculations at each of these sites is shown in Table 2.2.
The most extensive chemical database was at Dunham (Dunham II, Table 2.2), which
is one of the sampling sites for the UK Harmonised Monitoring Scheme (Littlewood *et
al.*, 1997; Robson and Neal, 1997). Trace metal concentration data are provided for
"dissolved" and "total acid-available" fractions. The "dissolved" fraction corresponds
to the fraction passing through a 0.45 μm membrane filter which was subsequently
acidified to 1% vv concentrated aristar nitric acid. The total acid-available fraction
was determined by acidifying an unfiltered sample (1% vv aristar nitric acid) and
agitating for 24 h at room temperature prior to filtration through 0.45 μm membrane
(see Neal *et al.*, 1997). River flows used in the calculation of load estimates were
determined from records collected at the flow gauging station at North Muskham. No
information was available for residual flows associated with tidal effects within the
tidal reaches of the Trent.

VARIABILITY OF RIVERINE CHEMICAL CONCENTRATIONS IN THE TIDAL REACHES

PARCOM recommend that the sampling site for load estimation should be in a region
of unidirectional freshwater flow (i.e. non-tidal). However, in practice, the location of

sampling is more variable (Jarvie *et al.*, 1997). For example, in the UK sampling is undertaken in the freshwater non-tidal reach as recommended by PARCOM, while in Germany and The Netherlands samples are taken from the tidal freshwater reach and in Belgium samples are collected within the saline tidal reach (Table 2.1). Comparison of pollutant load estimates based on sampling at different locations within tidal and non-tidal reaches may be problematic. Concentrations of chemical species are subject to change as they move through the tidal reaches, even given constant upstream inputs. For example, tides can affect chemical concentrations by providing energy for mixing of fresh and saline waters, transport of sediments and associated pollutants and moving water bodies back and forth (Ackroyd *et al.*, 1986; Morris *et al.*, 1982a,b). Mixing and dispersion effects in the tidal reaches are further modified by non-conservative behaviour of certain dissolved chemical constituents, by means of chemical reactions, adsorption/desorption from sediments, biological uptake, and release and transfer of materials across the benthic interface (Burton and Liss, 1976; Burton, 1988; Duinker, 1985). Non-conservative behaviour has been particularly well documented in tidal river reaches and estuaries as a result of changes in the physico-chemical properties of the water such as increases in ionic strength with mixing of fresh and saline waters and variations in pH (e.g. Morris, 1978; Morris and Bale, 1979; Morris, 1986; Boughriet *et al.*, 1992). However, "apparent" non-conservative mixing relationships may also occur as a result of point source or tributary inputs as well as temporal variations in river water chemistry which are shorter than the hydrodynamic residence time of water within the tidal river reach (Burton, 1988).

Before proceeding to riverine mass load estimation, an initial qualitative analysis of variations in chemical concentrations in the river was undertaken by plotting determinand–chloride relationships for all the samples collected within the tidal reaches (at Dunham, Gainsborough and Keadby). Chloride concentration is used as a conservative index of seawater–freshwater mixing. This analysis revealed that the highest degree of variability in determinand concentrations occurred under freshwater/very low salinity conditions. This is illustrated for five representative determinands: total oxidised nitrogen, suspended sediment, ammonia, and nickel in dissolved and total acid-available fractions (Figure 2.2). The high degree of scatter in the determinand–chloride mixing relationships is indicative of non-conservative behaviour, relating to sediment transport processes and changes in the speciation of dissolved components. The high variability in concentrations and non-conservative behaviour of determinands has important implications for load estimation in the tidal reaches, which are discussed later.

THE EFFECTS OF SAMPLING LOCATION ON LOAD ESTIMATION

To illustrate the effects of sampling location, mean annual loads of three contrasting determinands – total oxidised nitrogen (TON), suspended sediment and ammonia (NH_3) – were calculated for each of the four sites along the lower Trent (Winthorpe Bridge, Dunham, Gainsborough, Keadby) (Table 2.3). For this, PARCOM method 1 (Table 2.1) was applied to data collected between January 1986 and November 1988, using corresponding mean daily flow data from North Muskham.

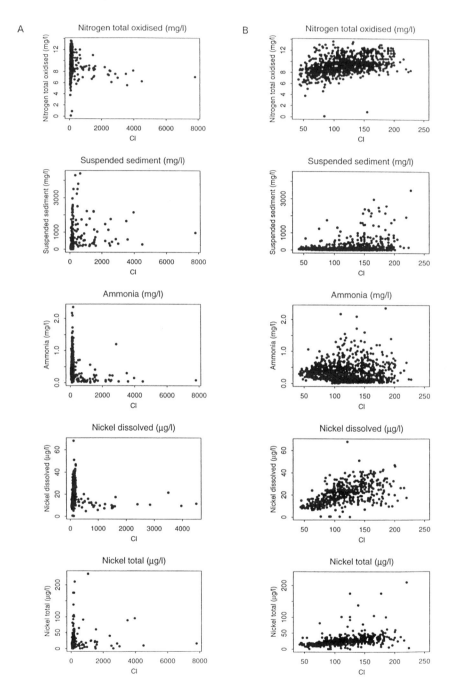

Figure 2.2 *Changes in concentrations of total oxidised nitrogen, suspended sediment, ammonia and dissolved and total acid-available nickel with chloride concentrations within the tidal River Trent. (A) Full data set; (B) for water samples with a chloride concentration less than 250 mg/l Cl⁻*

Table 2.3 *Mean annual loads of total oxidised nitrogen (TON), suspended sediment (SS) and ammonia (NH₃) along the lower reaches of the River Trent*

Sampling site (see Table 2.2)	Sampling period	Regime	Load SS (kt/a)	Load TON (kt/a)	Load NH₃ (kt/a)
Winthorpe Br I	26.2.86 to 9.11.88	Non-tidal	286	26.1	1.90
Dunham I	6.1.86 to 5.11.88	Tidal freshwater	92.8	28.4	1.59
Gainsborough I	6.1.86 to 15.11.88	Tidal freshwater	182	29.8	1.33
Keadby I	6.1.86 to 9.6.88	Tidal saline	339	28.1	0.517

The suspended sediment load at the non-tidal site, Winthorpe Bridge, is 286 kt/a; this decreases to 93 kt/a at the highest tidal site at Dunham and increases to 339 kt/a at the lowest tidal site at Keadby. Increases in suspended sediment loads and concentrations in the lower tidal reaches result from tidally driven physical processes. The freshwater discharge generates a residual seaward flow in the upper water of the tidal river, entraining seawater from the lower layer, which is provided by a residual landward flow closer to the river bed (Salomons and Forstner, 1984). This lower current is capable of mobilising the sediment trapped on the river bed and retaining sediment in suspension, which may subsequently be entrained into the surface water. The location of highest sediment loading (the "turbidity maximum") occurs in the region of maximum intrusion of seawater, which varies with tidal condition and river flow (Biggs and Cronin, 1981). Keadby (the tidal saline monitoring site) represents the closest of the sampling sites to the average location of maximum saline intrusion during the year, and thus exhibits the highest sediment loads. The sediment load at Winthorpe Bridge is unexpectedly high compared with Dunham and Gainsborough. This probably results from local availability of particulate sources or differences in sampling strategy, possibly relating to sampling high flow events.

TON concentrations are slightly enhanced in the tidal reaches relative to the non-tidal site, although this is no more than a 14% increase. The increase within the tidal reaches may reflect nitrification of the organic-rich near-bottom waters and mobilisation of this organic material during tidal events (e.g. Zhang, 1996). Ammonia loads demonstrate a systematic decrease downstream from the non-tidal into the tidal reaches. This suggests uptake of ammonia, possibly by adsorption to sediments, biological uptake or nitrification to nitrite or nitrate (therefore becoming part of the total oxidised nitrogen load).

THE EFFECTS OF SAMPLING UNDER DIFFERENT TIDAL CONDITIONS

In some cases, PARCOM contracting parties target sampling towards particular tidal conditions (Table 2.1). For example, in The Netherlands, sampling is undertaken at low tide (slack low water), and in Germany, sampling is undertaken on the ebb tide. Table 2.4 shows mean maximum and minimum concentrations, standard deviation and calculated mean annual loads of total oxidised nitrogen, ammonia and suspended sediment for data collected in a study to look at the effects of tide conditions on water quality. Data are available for flow tide, ebb tide, slack high water and slack low water conditions for the River Trent at Gainsborough and Keadby.

Table 2.4   *The effects of sampling under different tidal conditions on concentrations and loads of suspended sediment, total oxidised nitrogen and ammonia*

Determinand	Site	Tidal condition	Mean conc. (mg/l)	SD (mg/l)	Max. (mg/l)	Min. (mg/l)	Mean annual load (kt/a)
Suspended sediment	Gainsborough	Flow tide	299	431	2160	19	736
		Ebb tide	146	287	2150	3	254
		Slack high water	148	213	850	15	259
		Slack low water	159	163	720	10	338
	Keadby	Flow tide	1020	982	3800	38	1872
		Ebb tide	700	891	4400	24	1323
		Slack high water	411	450	2210	33	873
		Slack low water	412	660	3500	14	735
Total oxidised nitrogen	Gainsborough	Flow tide	9.73	1.34	12.5	6.95	25.6
		Ebb tide	9.36	1.53	13	5.03	22.7
		Slack high water	9.55	1.39	12.5	7.01	24.7
		Slack low water	9.34	1.56	13	3.76	23.4
	Keadby	Flow tide	8.98	1.69	12.5	5.56	21.8
		Ebb tide	9.48	1.59	13.6	6.03	24.0
		Slack high water	8.9	1.49	12.5	6.13	23.6
		Slack low water	9.51	1.93	13.5	0.1	23.8
Ammonia	Gainsborough	Flow tide	0.24	0.24	0.9	0.03	0.69
		Ebb tide	0.3	0.3	2.35	0.04	0.95
		Slack high water	0.26	0.17	0.74	0.03	0.84
		Slack low water	0.28	0.22	1.15	0.03	0.81
	Keadby	Flow tide	0.13	0.18	0.65	0.04	0.75
		Ebb tide	0.33	0.26	1.22	0.03	1.1
		Slack high water	0.26	0.19	0.74	0.03	0.84
		Slack low water	0.27	0.21	1.2	0.03	0.84

Highest sediment loadings are found under flow tide conditions at both sites. This is because during the flow tide, the residual bottom layer landward flow and tidal scour are highest, resulting in remobilisation of sediment deposited during the ebb flow and slack water conditions. At Gainsborough, highest TON loads occur under flow tide and high water conditions. This may reflect release of nitrate and nitrite produced by nitrification within organic-rich sediments, during mobilisation of bed material by tidal scour. In contrast, at Keadby, the highest TON loads are found during the ebb tide. This may indicate localised point sources of nitrate/nitrite, which would be dominant under ebb and slack water conditions. Ammonia loads are highest under ebb flow conditions at both sites. This might result from (a) removal of ammonia by adsorption to sediments mobilised during the flow tide or (b) dilution of ammonia in river water with relatively low-ammonia estuarine water.

THE EFFECTS OF BIASING SAMPLING TOWARD PARTICULAR RIVER FLOW CONDITIONS

PARCOM suggest that sampling for river load estimation should "aim to cover the whole flow cycle, but should concentrate on periods of expected high river flow". However, only the UK report an attempt to bias sampling towards high river flow (OSPARCOM, 1995) (Table 2.1).

To examine the effects of sampling high river flows on riverine mass loads, a comparison was made for a selection of key determinands between (i) mean annual mass loads calculated using the full concentration and flow data set, and (ii) mean annual mass loads calculated using the same dataset, but excluding samples collected during the highest 10% of river flows (Table 2.5). This analysis was undertaken at Dunham, where the most extensive chemical database was available (Table 2.2). The determinands were chosen to include a range of trace metals in dissolved and total acid-available phases (Fe, Pb, Cd, Zn, Cu, Ni), nutrients in the dissolved phase (TON and NH_3) and suspended sediment.

Considering PARCOM load estimation method 1, most determinand loads are lower when samples collected within the top 10% of flows are excluded. For many determinands, the effects are limited. However, there are cases where substantial effects are seen. Suspended sediment, Fe and total acid-available Pb show a marked decrease in loads estimated when the highest flows are excluded. This is because sediment loads and sediment-associated determinands tend to increase significantly at higher flows. Exclusion of the highest flows thus leads to significant under-estimation. For dissolved and acid-available nickel the opposite effect is seen, with a marked increase in loads estimated from the lower flows. This presumably reflects the importance of point source contributions of nickel. Point sources are diluted at high flows, so exclusion of high flow points leads to an over-estimate of the load.

The results shown here indicate that, for the two PARCOM load estimation methods considered here, it is extremely important that the sampling strategy should provide representative sampling of the flow regime. Whilst it is valuable to ensure that high flows are sufficiently well represented, over-sampling of high flows may also lead to error.

Table 2.5 *Mean annual mass loads at Dunham II (see Table 2.2), using PARCOM methods 1 and 2 (see Table 2.1), for (a) the full data set and (b) excluding the highest 10% of river flows*

Determinand	Method 1		Method 2	
	(a) Load using all data	(b) Load excluding the highest 10% of river flows	(a) Load using all data	(b) Load excluding the highest 10% of river flows
TON (kt/a)	23.4	25.3	22.7	18
SS (kt/a)	93.1	57.1	89.8	42.6
NH₃ (kt/a)	0.93	0.92	0.90	0.69
Fe (dis) (t/a)	209	104	205	117
Fe (tot) (t/a)	1650	1062	1612	1085
Pb (dis)(t/a)	7.0	6.4	6.5	5.1
Pb (tot)(t/a)	30.7	23.8	27.5	19.5
Cd (dis)(t/a)	0.68	0.72	0.63	0.56
Cd (tot)(t/a)	1.0	0.98	0.91	0.75
Zn (dis)(t/a)	93.3	92.9	88.6	74.9
Zn (tot)(t/a)	137	121	129	94.3
Cu (dis)(t/a)	24.8	25.4	24.0	19.6
Cu (tot)(t/a)	35.7	33.5	32.0	26.5
Ni (dis)(t/a)	47.2	61.2	44.3	41.5
Ni (tot)(t/a)	54.8	68.3	49.9	46

TON, total oxidised nitrogen; SS, suspended sediment; dis, dissolved load; tot, total acid-available load

THE EFFECTS OF LOAD ESTIMATION METHOD

PARCOM recommend that the load of a specific substance transported by a river should be estimated by taking the product of the mean flow-weighted concentration and the long-term average flow for the full sampling period (Table 2.2, PARCOM method 1). In those cases where insufficient flow data are available to use method 1, the pollutant load should be estimated by taking the average of the product of flow and concentration for a series of measurements (Table 2.2, PARCOM method 2). The effects of using these two algorithms are examined for the River Trent at Dunham, using the full data set and the same data set minus the highest 10% of flows (Table 2.5). Method 2 produces a lower estimate than method 1 in most cases. Essentially, method 2 and method 1 estimate loads by a mean flow-weighted concentration, scaled by a flow factor (Littlewood, 1992, 1995; Webb *et al.*, 1997). In method 1, the flow factor is the total long-term flow for the full sampling period, determined from an established gauging station. In method 2, the total annual flow is estimated using only the flows at the time of sampling. Thus, for method 2, if non-representative flows are sampled, the total annual flow may be considerably in error. This accounts for the PARCOM preference for use of method 1. If determinands have been sampled on the same dates, the ratio of method 2 to method 1 load estimates will be constant for those determinands.

Overall, method 2 gives similar results to method 1, provided that the flow regime is representatively sampled. If the flow regime is biased towards high or low flows, method 2 will usually provide inferior results.

THE EFFECTS OF SAMPLING FREQUENCY

PARCOM (1992) recommends that, for the major load-bearing rivers "there should be a minimum of 12 datasets within a 12-month period in order to estimate the annual input load … the sampling frequency may be increased beyond the minimum 12 datasets for those rivers carrying the heaviest contaminant loads. With such rivers, it should not be necessary to take samples more than once per week".

The data set from Dunham was used to conduct a brief evaluation of the effects of sampling frequency on pollutant mass load estimates. The mean sampling interval was 7.12 days over a period of 6.7 years (see Table 2.2, Dunham II). The effect of sampling interval was considered for those determinands measured most frequently (TON, suspended sediment, dissolved Cu and total acid-available Pb, Cd, Zn, Cu and Ni). Loads were calculated using PARCOM method 1. Subsamples were selected using a computerised routine at seven, 14, 28 and 56 day intervals. The sampling interval affects the number of possibilities for load estimates because, as the sampling interval increases, the number of choices of data subsets increases. A sampling interval of seven days thus allowed a possible maximum of seven separate load estimates, whereas a sampling interval of 14 days allowed a maximum of 14 load estimates. The mean annual load estimate and the coefficient of variation (the standard deviation as a percentage of the mean) for each determinand and sampling frequency are shown in Table 2.6. The variability of load estimates increases with sampling interval, indicating greater imprecision of estimates as the frequency of sampling declines. This arises primarily because the estimate is based on fewer samples, although part of this effect may relate to the greater number of possible load estimates generated at larger sampling intervals and confirms previous work (e.g. Ferguson, 1987; Littlewood, 1992, 1995; Walling and Webb, 1985; Webb *et al.*, 1997). Variation is greatest for suspended sediment loads and for sediment-associated metal concentrations, i.e. for determinands which tend to increase in concentration at high flows. A single high flow event may have a profound effect on sediment-associated pollutant loads, and the probability of catching such an event decreases at low sampling frequencies. Determinands in solution (e.g. dissolved Cu and TON) exhibit much lower variations in load estimates. These determinands are generally less affected by changes in flow.

It appears that average annual load estimates for the River Trent at Dunham, based on a 6.7 year data record, are not markedly affected by sampling interval. Sampling interval has an important effect on variability of load estimates but little bias is incurred as a result of the frequency of sampling.

SUMMARY

A summary of the effects of applying different PARCOM sampling and load estimation methodologies on mean annual riverine loads is shown in Table 2.7, using

Table 2.6 Mean annual load estimates and coefficients of variation (CV: standard deviation as a percentage of the mean) for samples selected at 7, 14, 28 and 56 day intervals from the Dunham II data set (sampling period from 11.1.89 to 5.9.95) (see Table 2.2)

	Sampling interval (days)							
	7		14		28		56	
Determinand	Mean annual load	CV (%)	Mean annual load	CV (%)	Mean annual load	CV (%)	Mean annual load	CV (%)
TON (kt/a)	23.4	0.23	23.4	0.26	23.7	2.2	23.8	3.6
SS (kt/a)	93.5	2.8	96.0	8.1	94.1	16.0	92.7	18.7
NH₃ (kt/a)	0.94	0.41	0.94	2.4	0.94	3.9	0.95	4.6
Pb (total) (t/a)	29.7	5.1	30.6	6.4	30.9	11.9	30.1	16.4
Cd (total) (t/a)	1	2.4	1.0	2.8	1.0	3.5	1	7.3
Zn (total) (t/a)	139	1.1	140	3.6	140	7.9	140	9.9
Cu (dissolved) (t/a)	25.3	0.3	25.3	0.6	25.2	0.7	25.3	1.1
Cu (total) (t/a)	36.2	0.77	37.2	1.6	37.6	4	36.6	7.4
Ni (total) (t/a)	55.0	0.7	54.6	1.8	54.8	2.4	54.3	5.9

TON, total oxidised nitrogen; SS, suspended sediment

Table 2.7   *Summary of the effects of the different sampling and load estimation strategies on mean annual loads of suspended sediment (SS) and total oxidised nitrogen (TON) for the River Trent*

Sampling/load estimation strategy			Load SS (kt/a)	Load TON (kt/a)
(A) Location of sampling	Non-tidal (Winthorpe I)		286	26.1
	Tidal freshwater (Dunham I)		92.8	28.4
	Tidal freshwater (Gainsborough I)		182	29.8
	Tidal saline (Keadby I)		339	28.1
(B) Sampling at particular tidal conditions	Tidal freshwater (Gainsborough II, III and V)	Flow tide	736	25.6
		Ebb tide	254	22.7
		Slack low water	338	23.4
	Tidal saline (Keadby II, III and V)	Flow tide	1872	21.8
		Ebb tide	1323	24.0
		Slack low water	735	23.8
(C) Sampling high river flows (Dunham II)	All data (including some high flows)		93.1	23.4
	Minus 10% highest flows		57.1	25.3
(D) Use of two load estimation methods	All data, Method 1		93.1	23.4
	All data, Method 2		89.8	22.7
(E) Sampling interval (Dunham II)	7 day		93.5	23.4
	14 day		96.0	23.4
	28 day		94.1	23.7
	56 day		92.7	23.8

For sampling site identification in parentheses, see Table 2.2

suspended sediment and TON to illustrate the major differences between particulate and dissolved loads.

For suspended sediment loads, the most marked effects result from (1) the location of sampling in tidal and non-tidal reaches and (2) for tidal sites, the state of the tide at the time of sampling. Suspended sediment loads are very sensitive to differences in location of sampling between non-tidal and tidal reaches, owing to differences in tidal action and sediment supply. This resulted in large differences between load estimates at different locations, with an increase in loads from 93 kt/a to 339 kt/a between the freshwater tidal site at Dunham and saline tidal site at Keadby. Sampling under different tidal conditions also has an important effect on sediment load estimates: for example, at Gainsborough, the suspended sediment load is 254 kt/a at ebb tide, compared with 736 kt/a at flow tide. Including samples collected at high flows also has a significant effect. Sediment loads calculated from a sampling programme which includes high flow events were 39% higher than loads calculated from the same data set where samples corresponding to the top 10% of flows had been excluded. Sampling location, tides and high river flows have more limited effects on mass-loads of TON, as TON does not exhibit such complex non-conservative flow-related transport. Flow and ebb tides account for no more than 11% change in TON loads and excluding the samples collected at the 10% highest river flows increased the load estimate by 8.1%.

In contrast with the effects of location and biasing of sampling to different tide and river flow conditions, sampling interval and load estimation methodology have a relatively minor effect when mean annual riverine loads are calculated for the River Trent using a 6.7 year data set. Changing the sampling interval from seven to 56 days causes a 0.8% reduction in estimated mean annual sediment load and a 0.1% increase in mean annual TON load. The use of PARCOM method 2 instead of method 1 results in a 3.5% reduction in mean annual sediment load and a 3.0% reduction in the mean annual TON load.

CONCLUSIONS

At the Fourth North Sea Conference held in Esbjerg in 1995, it was acknowledged that a major problem in assessing the success of policies to reduce pollutant inputs via rivers and estuaries has been "a lack of common, harmonised methodologies on which to calculate discharges from point and diffuse sources (including atmospheric emissions) as well as assessing inputs to the North Sea" (Andersen and Niilonen, 1995). At present, one of the key methodological discrepancies between PARCOM contracting parties has been the practice of sampling in tidal and non-tidal river reaches for estimation of riverine pollutant loads.

The results of the case study for the River Trent demonstrate that:

1. Sampling location and tidal and river flow conditions have a very marked effect on suspended sediment and particle-associated pollutant loads, and a lesser effect on dissolved determinand loads.
2. Sampling interval and load estimation method have a comparatively minor effect on mean annual sediment and chemical loads, compared with the location of sampling, sampling under different tidal conditions and the effects of sampling high river flows. However, should these data be used for a time series of annual loads, the frequency of sampling and load estimation method would assume much greater importance (Webb et al., 1997).

Therefore, the case study of the River Trent indicates that the present diversity in sampling strategies for load estimation by the PARCOM contracting parties at present fails to provide an effective basis for comparison of riverine mass loads at the European regional scale. To facilitate more meaningful comparisons of loads estimated from different river systems, improved understanding of the sensitivity of load estimates to sampling and estimation methodologies is urgently required for the other major river systems draining into the North Sea. The dynamics of water quality in any given tidal river system depends on site-specific physical and chemical factors. These include freshwater runoff, residence times of water within the tidal reaches, the nature and sources of pollutants and factors affecting the speciation of chemical determinands such as salinity, pH, availability of complexing species, the characteristics and composition of the suspended sediment, and biological interactions. An examination of the implications of water quality dynamics for load estimation requires intensive measurement programmes, site-specific empirical studies and improved process understanding.

The key to future success in understanding the nature and magnitude of riverine mass loads to coastal waters ultimately lies in the success of linking within-river processes and dynamic modelling studies (e.g. Lewis *et al.*, 1997; Whitehead *et al.*, 1997) with estuarine process-based dynamic water quality and flow modelling (e.g. Falconer and Owens, 1990; Harris *et al.*, 1984, 1991; Ng *et al.*, 1996). This coupling of river and estuarine models is a major goal of the UK Natural Environment Research Council's Land–Ocean Interaction Study (Wilkinson *et al.*, 1997) and it is hoped that this will pave the way toward predicting the propagation of riverine chemical fluxes through the tidal reaches, estuary and coastal zone.

ACKNOWLEDGEMENTS

The Environment Agency is thanked for providing access to the Lower Trent chemistry and river flow data. This work was carried out under the Land–Ocean Interaction Study (LOIS), funded by the Natural Environment Research Council.

REFERENCES

Ackroyd, D.R., Bale, A.J., Howland, R.J.M., Knox, S., Millward, G.E. and Morris, A.W. 1986. Distribution of dissolved Cu, Zn and Mn in the Tamar estuary. *Estuarine, Coastal and Shelf Science*, **23**(5), 621–640.

Andersen, J. And Niilonen, T. 1995. *Fourth International Conference on the Protection of the North Sea*, Progress Report, Esbjerg, Denmark, 8–9 June 1995. Ministry of Environment and Energy, Danish Environmental Protection Agency, Stradgade 29, DK1401, Copenhagen.

Biggs, R.B and Cronin, L.E. 1981. Special characteristics of estuaries. In Neilson, B.J. and Cronin, L.E. (eds), *Estuaries and Nutrients*. Humana Press, New Jersey.

Bougriet, A., Ouddane, B., Fischer, J.C., Wartel, M. and Leman, G. 1992. Variability of dissolved Mn and Zn in the Seine estuary and chemical speciation of these metals in suspended matter. *Water Research*, **26**(10), 1359–1378.

Burton, J.D. 1988. Riverborne materials at the continent-ocean interface. In Lerman, A. and Meybeck, M. (Eds), *Physical and Chemical Weathering in Geochemical Cycles*. Kluwer, Dordrecht, 299–321.

Burton, J.D. and Liss, P.S. 1976. *Estuarine Chemistry*. Academic Press, London.

De Vries, A. and Klavers, H.C. 1994. Riverine fluxes of pollutants: monitoring strategy first, calculation methods second. *European Water Pollution Control*, **4**(2), 12–17.

Dickinson, W.T. 1981. Accuracy and precision of suspended sediment loads. In *Erosion and Sediment Transport Measurement*, Proceedings of Florence Symposium, June 1981. IAHS Publication No. 133, 195–211.

Duinker, J.C. 1985. Estuarine processes and riverborne pollutants. In: Nurnberg, H.W. (Ed.), *Pollutants and their Ecotoxicological Significance*. John Wiley, Chichester, 227–238.

Falconer, R.A. and Owens, P.H. 1990. Numerical modelling of suspended sediment fluxes in estuarine waters. *Estuarine, Coastal and Shelf Science*, **31**, 745–762.

Ferguson, R.I. 1987. Accuracy and precision of methods for estimating river loads. *Earth Surface Processes and Landforms*, **12**, 95–104.

Harris, J.R.W., Bale, A.J., Bain, B.L., Mantoura, R.F.C., Morris, A.W., Nelson, L.A., Radford, P.J., Uncles, R.J., Weston, F.A. and Widdows, J. 1984. A preliminary model of the dispersion and biological effect of toxins in the Tamar estuary, England. *Ecological Modelling*, **22**, 253–284.

Harris, J.R.W, Hamblin, C.C. and Stebbing, A.R.D. 1991. A simulation study of the effectiveness of legislation and improved dockyard practice in reducing TBT concentrations in the Tamar estuary. *Marine Environment Research*, **32**, 279–292.

Jarvie, H.P., Neal, C. and Tappin, A.D. 1997. European land-based pollutant loads to the North Sea: an analysis of the Paris Commission data and review of monitoring strategies. *Science of the Total Environment*, **194/195**, 39–58.

Lewis, D.R., Williams, R.J. and Whitehead, P.G. 1997. Quality simulation along rivers (QUASAR): an application to the Yorkshire Ouse. *Science of the Total Environment*, **194/195**, 399–418.

Littlewood, I.G. 1992. *Estimating Contaminant Loads in Rivers: a Review*. Report 117, Institute of Hydrology, Wallingford, 175pp.

Littlewood, I.G. 1995. Hydrological regimes, sampling strategies and assessment of errors in mass load estimates for United Kingdom rivers. *Environment International*, **21**(2), 211–220.

Littlewood, I.G., Watts, C.D., Green, S., Marsh, T.J. and Leeks, G.J.L. 1997. *Aggregated river mass loads for Harmonised Monitoring Scheme catchments grouped by PARCOM coastal zones around Britain*. Report to the Department of the Environment (EPG 1/8/26), Institute of Hydrology, Wallingford.

Morris, A.W. 1978. Chemical processes in estuaries: the importance of pH and its variability. In *Environmental Biogeochemistry and Geomicrobiology*, Volume 1. *The Aquatic Environment*. Ann Arbor Science Publishers, Michigan, 179–187.

Morris, A.W. 1986. Removal of trace metals in the very low salinity region of the Tamar estuary, England. *Science of the Total Environment*, **49**, 297–304.

Morris, A.W. and Bale, A.J. 1979. Effect of rapid precipitation of dissolved Mn in river water on estuarine Mn distributions. *Nature*, **279**(5711), 318–319.

Morris, A.W., Mantoura, R.F.C., Bale, A.J. and Howland, R.J.M. 1978. Very low salinity regions of estuaries: important sites for chemical and biological reactions. *Nature*, **274**(5672), 678–680.

Morris, A.W., Bale, A.J. and Howland, R.J.M. 1982a. The dynamics of estuarine manganese cycling. *Estuarine, Coastal and Shelf Science*, **14**(2), 175–192.

Morris, A.W., Bale, A.J. and Howland, R.J.M. 1982b. Chemical variability in the Tamar estuary, south-west England. *Estuarine, Coastal and Shelf Science*, **14**(6), 649–661.

Neal, C., Robson, A.J., Jeffery, H.A., Harrow, M.L., Neal, M., Smith, C.J. and Jarvie, H.P. 1997. Trace element inter-relationships for the Humber rivers: inferences for hydrological and chemical controls. *Science of the Total Environment*, **194/195**, 321–343.

Ng, B., Turner, A., Tyler, A.O., Falconer, R.A. and Millward, G.E. 1996. Modelling contaminant geochemistry in estuaries. *Water Research*, **30**(1), 63–74.

OSPARCOM (Oslo and Paris Commissions). 1995. *Data report on riverine and direct inputs of contaminants to the waters of the Paris Convention in 1993*. Oslo and Paris Commissions, London.

PARCOM (Paris Commission). 1992. *Principles of the comprehensive study of riverine inputs*. Oslo and Paris Commissions, London.

Robson, A.J. and Neal, C. 1997. A summary of the regional water quality for the Eastern UK rivers. *Science of the Total Environment*, **194/195**, 15–37.

Salomons, W. and Forstner, U. 1984. *Metals in the Hydrocycle*. Springer Verlag, Heidelberg.

Walling, D.E. and Webb, B.W. 1981. The reliability of suspended sediment load data. In *Erosion and Sediment Transport Measurement*, Proceedings of Florence Symposium, June 1981, IAHS Publication No. 133, 177–194.

Walling, D.E. and Webb, B.W. 1985. Estimating the discharge of contaminants to coastal waters by rivers: some cautionary comments. *Marine Pollution Bulletin*, **16**, 488–492.

Walling, D.E. and Webb, B.W. 1988. The reliability of rating curve estimates of suspended sediment yield: some further comments. In Bordas, M.P. and Walling, D.E. (eds), *Sediment Budgets*, Proceedings of the Porto Alegre Symposium, December 1988, IAHS Publication No. 174, 337–350.

Webb, B.W., Phillips, J.M., Walling, D.E., Littlewood, I.G., Watts, C.D. and Leeks, G. 1997. Load estimation methodologies for British Rivers and their relevance to the LOIS RACS(R) programme. *Science of the Total Environment*, **194/195**, 379–389.

Whitehead, P.G., Williams, R.J. and Lewis, D.R. 1997. Quality simulation along river systems (QUASAR): model theory and development. *Science of the Total Environment*, **194/195**, 447–456.

Wilkinson, W.B., Leeks, A., Morris, A.W. and Walling, D.E. 1997. Rivers and coastal research in the Land Ocean Interaction Study. *Science of the Total Environment*, **194/195**, 5–14.

Zhang, J. 1996. Nutrient elements in large Chinese estuaries. *Continental Shelf Research*, **16**(8), 1023–1045.

3

Changing Standards and Catchment Sources of Faecal Indicators in Nearshore Bathing Waters

*D. Kay, M.D. Wyer, J. Crowther, J.G. O'Neill, G. Jackson,
J.M. Fleisher and L. Fewtrell*

BACKGROUND

Bathing water quality standards in European Union (EU) member countries are defined in the 1976 Bathing Water Directive (76/160/EEC) (Anon., 1976). The Directive applies to water where bathing is explicitly authorised or traditionally practised by "large numbers" of bathers (Article 1) and such locations are identified, or designated, by member countries which are then responsible for the implementation of a sampling programme designed to acquire data defining the water quality at any given site. Some 19 water quality parameters are listed in the Directive's Annex. However, compliance and/or non-compliance is generally assessed in terms of the coliform bacterial parameters (i.e. total and faecal coliform concentrations per 100 ml of seawater). The Directive defines mandatory (termed *Imperative*) and recommended (termed *Guide*) values for the concentration of faecal coliform (FC) and total coliform (TC) organisms. The *Imperative* standard requires 95% of samples collected through the bathing season to have $< 10\,000$ TC/100 ml and < 2000 FC/100 ml, respectively. The *Guide* standard requires that 80% of samples have < 500 TC/100 ml and < 100 FC/100 ml. In addition, *Guide* level compliance requires that 90% of samples should have < 100 faecal streptococci (FS)/100 ml. However, all EU member states do not measure FS concentrations at sufficient intensity to facilitate its uniform use in compliance assess-

Water Quality: Processes and Policy. Edited by Stephen T. Trudgill, Des E. Walling and Bruce W. Webb.
© 1999 John Wiley & Sons Ltd.

ment across the community (Kay *et al.*, 1996). Hence, the PASS/FAIL status of a bathing water is "generally" based on the coliform organism concentrations alone. Two further microbiological parameters are specified in the Directive: namely, enterovirus/10 litres and *Salmonella*/litre. The Directive requires these determinands to be absent for compliance. However, these parameters are not assessed by all EU member states and are thus not used in Community-wide compliance assessment.

In the mid-1970s, the Bathing Water Directive represented a departure from traditional practice for many EU member states in that it established the principle of a standard for a receiving water rather than setting a consent, or allowed concentration, for a polluting effluent stream. This has led to some problems of implementation because the principal management interventions for improving the quality of the environment have been either physical removal of point source pollution through the construction of a long outfall discharging at some point distant from the recreational water compliance points, and/or some form of treatment of the sewage flow. Treatment of the sewage stream involves some or all of the following processes:

- preliminary treatment of the sewage stream, involving masceration and/or screening, to enhance dispersion and bacterial die-off;
- primary treatment, involving settlement to remove particles and faecal aggregates;
- secondary, i.e. biological, treatment such as trickling filters or activated sludge;
- tertiary treatment involving disinfection by UV light, microfiltration or chemicals (e.g. chlorine and peracetic acid).

This suite of engineering interventions is technically feasible, familiar and highly effective in reducing the bacterial loading in the effluent stream. Sewage contains approximately 10^7 FC/100 ml. However, a good sewage treatment plant, employing secondary treatment with terminal UV disinfection, can achieve an effluent stream with < 100 FC/100 ml.

A major construction programme is underway in the UK to improve the quality of effluents discharged to coastal waters and/or remove the nearshore discharge locations offshore by the construction of long outfall pipes. In a recent report to a House of Lords Select Committee enquiry (Anon., 1994a, 1995a,b), Halcrow plc (Anon., 1995c) suggested that over £10 billion had been committed to achieve the requirements of the Bathing Water Directive and the Urban Waste Water Treatment Directive (91/271/ EEC) (Anon., 1991). This fiscally important expenditure has been authorised by the Environment Agency (EA) as regulator of the water companies who are responsible for sewage disposal and the implementation of enhanced sewage treatment. The authorisation of new schemes by the EA is vital to the water companies since EA approval guarantees that water companies can pass the costs of new infrastructure through to their customers. If the scheme is not EA approved, then the water company must bear the cost from existing operating revenues.

Thus, the main strategy employed to achieve compliance of UK beaches is point source control, generally involving expensive sewage treatment and disposal systems. However, a series of recent research studies suggests that this expenditure may not ensure compliance of UK bathing waters, for two reasons. First, the standards used to assess compliance are changing rapidly with developments at both EU (Anon., 1994b) and World Health Organisation (WHO) levels. Second, point source control of the sewage stream, used in isolation, ignores the other sources of faecal indicators discharged to coastal waters which can have highly significant implications for water quality compliance.

THE POLICY ENVIRONMENT

The existing Bathing Water Directive (76/160/EEC) establishes limits in terms of faecal indicators which have no basis in environmental epidemiology or public health science. In terms of international standards for bathing waters, the EU *Imperative* standard is amongst the least stringent whilst the *Guide* standard is perhaps the most stringent (Kay *et al.*, 1990). However, it is incorrect to claim that beaches passing either the *Imperative* or, indeed, the *Guide* standards are protective of public health. Public health protection is a stated aim of the Directive and this inability of the policy community to make scientifically supportable statements on the public health impacts of recreation in waters passing legally enforced standards has been a matter of concern for some time. This problem came to prominence following the work of Cabelli *et al.* (1982) which reported a clear dose–response relationship attributable to bathing in water with faecal coliform (*E. coli*) concentrations within both *Imperative* and *Guide* standards (Figure 3.1).

Cabelli *et al.* (1982) completed their studies in the United States at freshwater, brackish and marine beaches and, clearly, their work may not be directly transferable to an EU and UK context. However, their findings did point to a reconsideration of

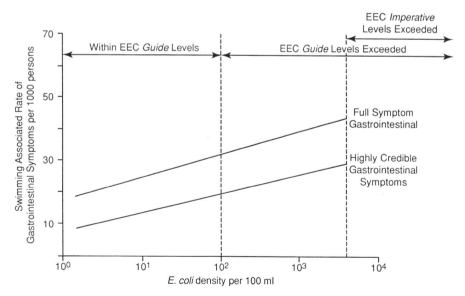

Figure 3.1 *Relation between swimming-related gastrointestinal symptoms and* E. coli *density. Adapted from Cabelli* et al. *(1982)*

the 1976 Directive standards. Interestingly, Cabelli *et al.* (1982) suggested a stronger link between concentrations of faecal streptococci and gastrointestinal symptoms following bathing in marine waters than with the faecal coliforms used for compliance assessment in Europe. These epidemiological dose–response relationships were subsequently used by the United States Environmental Protection Agency (USEPA, 1986; see also USEPA, 1976) to develop new health-related standards for US recreational waters which are still the subject of much discussion at the State level. This discussion has focused on certain epidemiological weaknesses in the research protocol adopted by Cabelli *et al.* (1982) which have been detailed by Fleisher (1990a,b, 1991, 1992) and Fleisher *et al.* (1993).

The effect of these US studies in Europe was to focus attention on the lack of epidemiological information to underpin the existing standards in Directive 76/160/EEC or, indeed, to suggest any new standard system which might have a firmer scientific foundation. This was highlighted in 1985 during the House of Commons Welsh Affairs Committee enquiry into coastal sewage disposal (Anon., 1985a,b,c). Here, the only UK epidemiology which could be used to influence the debate was the Public Health Laboratory Service/Medical Research Council study reported in 1959 (PHLS, 1959) which indicated that there might be little risk of contracting either poliomyelitis or enteric fever from recreational water exposure unless the waters were "so fouled as to be aesthetically revolting" (which was equated with a total coliform concentration of $> 10\,000/100\,ml$). This credible and carefully conducted retrospective case-control study was, however, restricted to two serious (i.e. life-threatening) illnesses, poliomyelitis and enteric fever. Less serious ailments shown to be associated with recreational water exposures by Stevenson (1953) were ignored.

Following the 1985 Commons Welsh Affairs Committee Enquiry, it was clear that a UK epidemiological investigation was required to underpin policy in this area and an expert advisory group was established by Government in 1987. The pitfalls in the US research protocol were known to this group largely through the work of Fleisher and his co-workers (Fleisher, 1990a,b, 1991, 1992; Fleisher *et al.* 1993). To avoid these problems in the subsequent UK and EU policy debate, an improved research protocol was designed which employed healthy adult volunteers who were randomised into bather and non-bather groups and then exposed to seawater passing existing EU *Imperative* standards during four weekend studies around the UK coast from 1989 to 1992 (Figure 3.2). This protocol was first described by the WHO in 1972 (WHO, 1972).

The results of these studies were reported in Kay *et al.* (1994) and Fleisher *et al.* (1996). The randomised volunteer protocol allowed greater control of (and information acquisition on) both non-water-related risk factors (e.g. food intake) and potential confounding factors (e.g. age and sex) for gastrointestinal (GI) illness. Perhaps more importantly, the protocol facilitated intensive sampling of water quality at the time and place of exposure. Thus, a specific index of bacteriological water quality could be assigned to each volunteer. The effect of this was to reduce the "misclassification bias" encountered in previous epidemiological investigations where a "daily mean" value was used as the exposure measure for all bathers at a beach for a particular day. Such allocation of an inherently imprecise exposure measure produces "misclassification bias" which causes a systematic, and substantial, under-estimate of the outcome variable (i.e. illness) (Fleisher, 1991; Fleisher *et al.*, 1993).

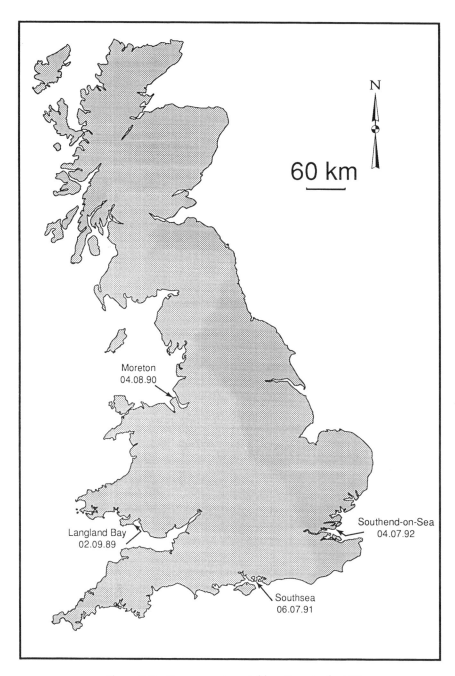

Figure 3.2   *Four experimental locations in the UK*

Table 3.1

Reported symptom	Most Predictive Indicator	Measurement Depth	Were Bathers at Greater Risk?
Gastroenteritus	Faecal streptococci	Chest	Yes
AFRI*	Faecal streptococci	Chest	Yes
Ear symptoms	Faecal coliform	Chest	Yes
Eye symptoms	None	N/A	Yes
Skin symptoms	None	N/A	No

*AFRI = Acute febrile respiratory illness

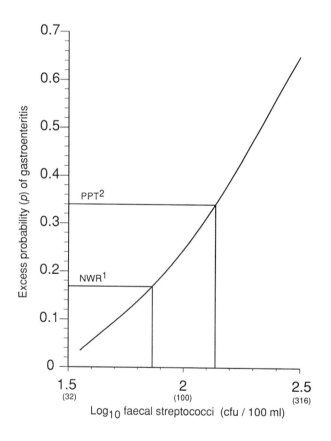

1 NWR = Non-water related risk level
(pGI = 0.17, FS = 73 cfu / 100 ml)

2 PPT= Person to Person Transmission
risk level (pGI = 0.34, FS = 137 cfu / 100 ml)

Figure 3.3   *Results of UK randomised healthy volunteer investigation. Adapted from Kay et al. (1994)*

Figure 3.3 and Table 3.1 summarise the results of the UK randomised healthy volunteer investigations. The main water quality parameter predicting gastroenteritis and upper respiratory tract illness was faecal streptococci (FS) measured at chest depth. This logistic regression relationship between FS concentration and GI illness is shown in Figure 3.3 which suggests a high and significant excess probability of gastroenteritis at faecal streptococci concentrations well within the *Guide* level for this parameter (i.e. 100/100 ml) specified in Directive 76/160/EEC.

Prior to the formal publication of these results in 1994, the European Commission produced a Proposal for an amendment to the Bathing Water Directive (Anon., 1994b) which suggested a new *Imperative* standard for faecal streptococci of 400/100 ml to be achieved in approximately 95% of samples. A House of Lords Select Committee gave detailed consideration to the Commission Proposal in 1994 and 1995 (Anon., 1994a, 1995a). Again, the lack of scientific data to underpin the proposed standard for faecal streptococci was noted. The Commission's view, expressed to the Lord's Committee, was that their proposal was broadly "compliance neutral". This stance might have been based on an examination of historical data which, at least for the UK, suggested that the 95 percentile point for the UK-wide faecal streptococci probability density function was just over 400/100 ml (Kay, 1994). However, it failed to consider the effect of the high, but imperfect, correlation between total coliform, faecal coliform and faecal streptococci bacteria. Where compliance against all three faecal indicator bacteria is required (rather than just the coliform parameters) under the new *Imperative* standard, then a significant reduction was observed in the numbers of PASS beaches when the proposed standards were applied to historical data. The effect would be to reduce the *Imperative* PASS rate from 80–90% to 60–70% for UK beaches (Anon., 1994a, 1995a).

Following comment by the Lord's Committee and authoritative UK witnesses, including the National Rivers Authority (NRA) (the NRA is now subsumed into the Environment Agency), the Commission initiated an external review of the scientific evidence which could be used to underpin an appropriate standard for faecal streptococci (Kay *et al.*, 1996; Kay and Wyer, 1997; Kay and Rees, 1997; Kay *et al.*, 1999). The report of this group suggested a novel approach to the definition of standards for recreational waters first developed for the States of Jersey (Wyer and Kay, 1995). In essence, the approach combines the faecal streptococci probability density function (p.d.f.) for any location with the epidemiological dose–response curve to produce an estimate of disease burden. Figures 3.4, 3.5 and 3.6 illustrate this process.

The p.d.f. used in this example approximates to the "average" UK beach using > 30 000 faecal streptococci determinations. The implication of this analysis is that if 1000 persons were exposed to UK seawater of this quality, then 679 would not encounter water that might make them ill. Some 321 would encounter water which might make them ill and, of these, 71 would exhibit symptoms of gastroenteritis.

Kay *et al.* (1996) used this generic approach to suggest a series of water quality standards to the Commission based on clearly defined "risk" levels.

These assume that standards will be defined in terms of 95% compliance levels, although a strong argument could be put forward to design standards in terms of the geometric mean faecal streptococci concentrations and its $\log_{10}$ standard deviation which can be used to provide an estimate of disease burden or risk (Kay *et al.*, 1999).

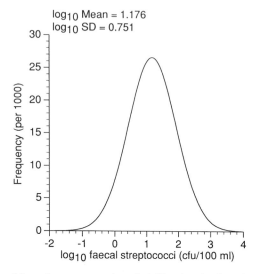

Figure 3.4   *Example of faecal streptococci probability density function. Adapted from Wyer and Kay (1995)*

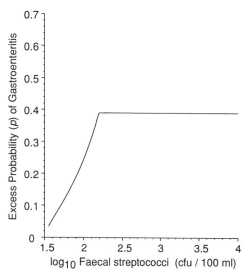

Figure 3.5   *Example of epidemiological dose–response curve. Adapted from Wyer and Kay (1995)*

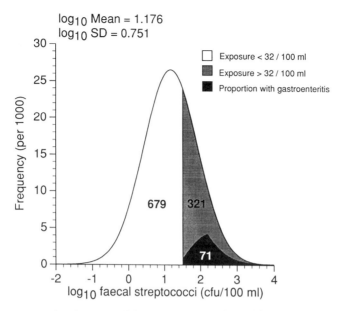

Figure 3.6 *Example of estimated disease burden. Adapted from Wyer and Kay (1995)*

This generic approach has been reviewed in a series of WHO consultation meetings (Bad Elster 1996, Jersey 1997) and was adopted to underpin new draft WHO Guidelines for recreational waters.

It is clear, therefore, that, at both EU and WHO levels, there is considerable potential for amended recreational water standards. These will almost certainly be more stringent and implemented in the next few years. The implication of this is that existing schemes designed to achieve point source control as a means of compliance with Directive 76/160/EEC may well be under-designed if the goal is compliance with the present *Imperative* standard. However, the problems for sewage treatment engineers do not stop at this uncomfortable fact.

NON-SEWAGE SOURCES OF FAECAL BACTERIA

At many coastal bathing locations, exceedence of bacteriological standards occurs following rainfall. This "rainfall effect" has been observed even following expensive and effective treatment of the sewage stream. Perhaps the best illustration of this effect was observed in the States of Jersey in the 1993 bathing season immediately following the installation of the first large UV plant in Europe. This system kills bacteria and viruses in the final effluent producing a microbiologically "clean" effluent (typically < 100 FC/100 ml). A monitoring point in the bathing water adjacent to the "clean" effluent failed to achieve *Imperative* level compliance in the bathing season following installation of the UV system. However, as examination of Figure 3.7 shows, the two occasions on which non-compliance was evident were both associated with rainfall events.

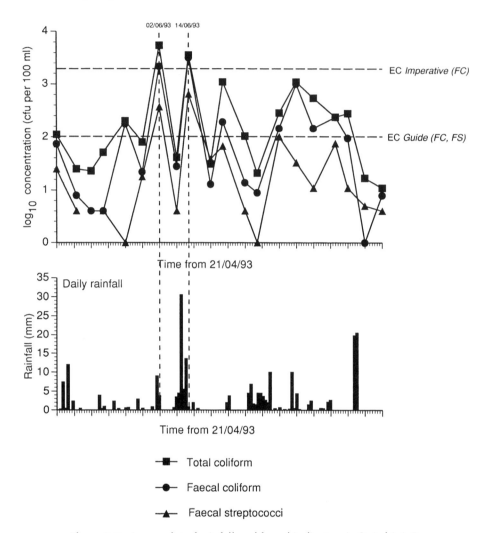

Figure 3.7   *Jersey plot of rainfall and faecal indicators in St Aubin's Bay*

This leads to the question of where the faecal bacteria associated with non-compliance following rainfall originated. There are three potential sources: first, high flows discharged from the sewerage system through storm overflows which are installed to relieve pressure and prevent backflow into domestic properties via the toilet system; second, partly treated effluent which is not fully disinfected by UV light during high flow events; and, finally, bacterial concentrations in streams and rivers draining into the coastal waters from adjacent catchment areas.

Engineering consideration of the delivery of faecal indicators to nearshore recreational waters at the scheme design stage has traditionally centred on the first two, because these elements are what the scheme seeks to manage. This might be an acceptable perspective if the objective of the scheme was to comply with a discharge "consent". However, it is clearly inappropriate where the scheme is designed to improve unacceptable quality in the receiving waters themselves. In these circumstances, all inputs should be examined. This process, however, raises some non-trivial questions of sampling programme design and implementation.

Clearly, the estimation of faecal bacterial delivery from streams draining catchment areas requires an understanding of the stream's microbial response to changes in flow. Very few studies worldwide have reported the microbial concentrations in a stream during the passage of individual hydrograph events (McDonald *et al.*, 1982; Wilkinson *et al.*, 1995; Wyer *et al.*, 1995, 1997a,b). However, the available evidence suggests that microbial concentrations exhibit a rapid increase during the hydrograph passage which, in the case of small rural streams, occurs during the rising limb of the hydrograph event. Concentrations during this "flush" of bacteria can exceed $10^6/100$ ml in streams not obviously affected by human sewage wastes. This represents a two to three order increase in concentration alone. When added to the volumetric (one to two order) increase associated with high flow events, then the clear potential importance of high flow episodes in total bacterial delivery to nearshore recreational waters becomes clear. If this conceptual model of faecal bacterial input is correct, then a highly episodic, pulsed input of faecal indicators to nearshore waters might be predicted. The interesting scientific and management question is whether this episodic input is more or less important than the delivery from the sewerage system during periods of high flow when compliance problems are commonly experienced.

A two year fieldwork programme was designed and implemented in the States of Jersey to address this question. It involved intensive sampling during high flow events which occurred in the bathing season and regular tri-weekly sampling to acquire data on baseflow water quality (Wyer *et al.*, 1997a). Delivery estimates were made using routine hydrological data from existing sites on the island.

The discharge from sources into St Aubin's Bay is shown in Figure 3.8a, whilst the delivery of faecal indicators is shown in Figure 3.8b. Clearly, flow from the sewage plant dominates but the plant makes only a minor contribution to bacterial delivery.

This is even more marked when the high flow data are considered. Here, stream high flows contribute some 77% of the total input to the Bay with UV disinfection installed at the sewage treatment plant. Figure 3.9 illustrates these contributions assuming a sewage input without and with terminal UV disinfection.

The Jersey data provided the first quantification of the range of faecal indicator sources available in the UK. This problem had been evident following the installation of the UV plant and this made the Jersey case study unique and potentially unrepresentative. However, the problem of beach non-compliance following rainfall is common at UK and EU bathing waters where drainage is to the coast. An opportunity to acquire data from a possibly more representative UK bathing water at Staithes on the North Yorkshire coast became available in 1996. Here, sewage is discharged without treatment. A similar approach to that developed in Jersey was adopted, namely, the characterisation of high and low flow water quality, together with parallel hydrology to facilitate bacterial budget estimates.

The Staithes budget of faecal coliform delivery is shown in Figure 3.10. This site is, again, affected by non-compliance after rainfall-induced high flow events. During such episodes it is evident that some 68% of the faecal bacterial delivery is from the stream.

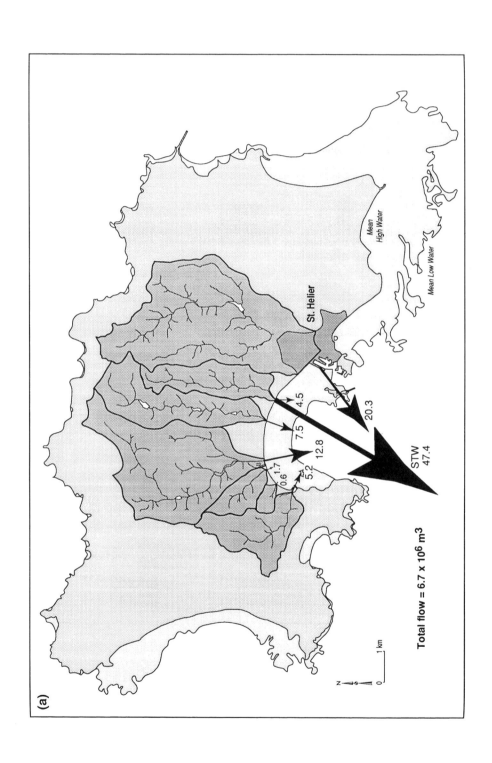

(a)

St. Helier

Mean High Water

Mean Low Water

4.5

7.5

12.8

1.7

0.6

5.2

20.3

STW 47.4

N

0   1 km

Total flow = 6.7 × 10⁶ m³

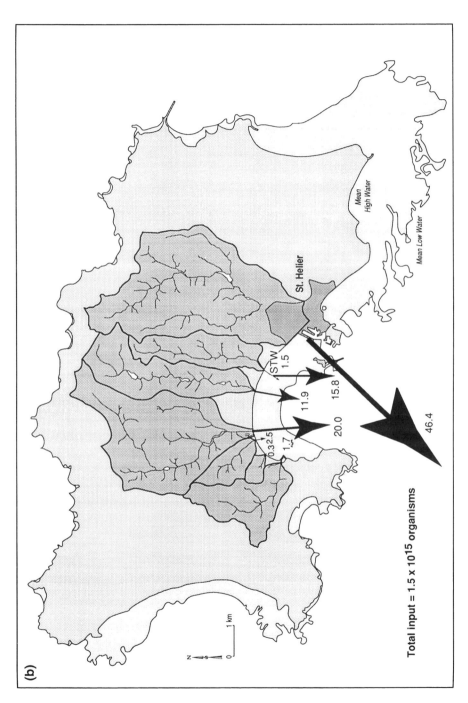

Figure 3.8 (a) Discharge from sources into St Aubin's. (b) Delivery of faecal indicators. From Kay et al. (1997a)

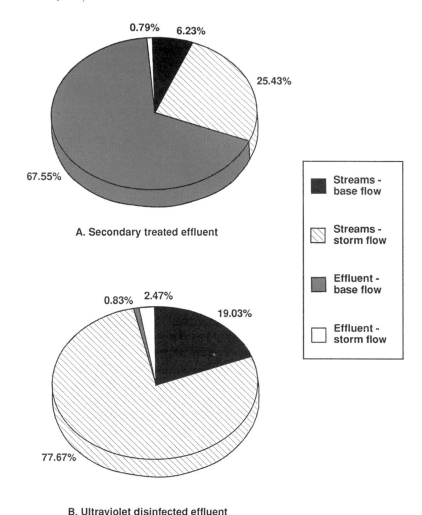

Figure 3.9 *Percentage contribution of stream and sewage effluent sources to faecal coliform loading in St Aubin's Bay, Jersey*

This is a surprising result given that the sewage stream receives no treatment to reduce its bacterial concentration. Clearly, enhanced sewage treatment would have an impact on baseflow inputs to the nearshore recreational waters but might have little impact on non-compliance after rainfall.

These are significant results for the designers of sewage treatment schemes. They clearly indicate that a total budget appraisal is required for all new schemes if post-commissioning non-compliance is to be avoided or at least predicted. Furthermore, they indicate that compliance of bathing waters may be more dependent on management and remediation of bacterial delivery from non-sewage sources, such as catchment streams, than from the sewerage system itself. Remediation of diffuse and

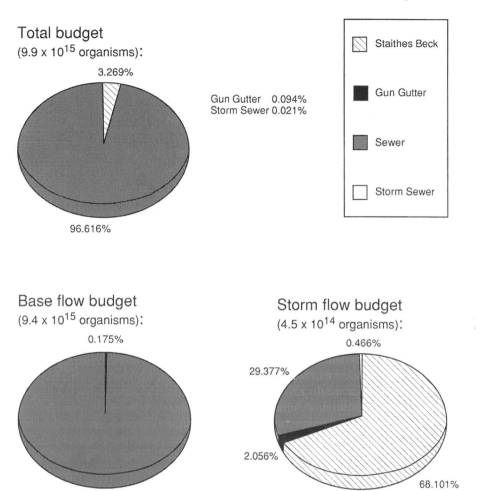

Figure 3.10 *Staithes budget of faecal coliform organism delivery*

non-point catchment sources of faecal indicators presents a new and complex series of challenges. A range of potential sources are evident, e.g. livestock on fields, deliberately spread livestock slurry and sewage sludge, farmyard washing and leakage from slurry stores, domestic properties having septic tanks or soakaways or cross-connections in their wastewater-disposal systems, sewage treatment plants discharging to rivers, diffuse flows from broken and/or blocked sewers and other polluting activities. Clearly, consideration of this range of sources with a view to remediation of non-sewage inputs to recreational waters requires an integrated assessment at the catchment scale.

Catchment modelling offers one mechanism for the prediction of diffuse sources of pollution derived mainly from livestock. Regression models have now been calibrated for the Jersey and Staithes catchments, with further work completed at Welsh study sites centred on the Nyfer and Ogwr catchments. The objective of this ongoing work is to produce a generic catchment model able to utilise available data to predict high and low flow delivery of faecal indicators to coastal waters, to underpin the initial appraisal or budget estimate of indicator sources to be used at the scheme design stage.

CONCLUSIONS

There are two directly related and potentially conflicting trends reported in this chapter. The first is the probable redesign of quality standards for recreational waters. Initial signs of this are seen in the EU Commission Proposal for tightened standards for recreational waters. There will, clearly, be political opposition to any such movement on cost grounds, but the fact that the WHO is also now moving in the direction of new Guidelines for recreational waters provides further strong evidence that, in the medium term at least, recreational water standards are likely to become more stringent. Epidemiological investigations undertaken in the UK have proven central to the debates both inside the EU and WHO which have surrounded this policy re-evaluation.

As these "goal posts" are moving for sewage treatment design engineers, there is an increased realisation that treatment of the sewage stream alone may not be sufficient to achieve even the present legislation, let alone more stringent "risk based" water quality standards now in the pipeline. The way forward must clearly embrace an integrated approach to the management of water quality in nearshore recreational waters. This should assess and quantify all faecal indicator inputs from both sewage and non-sewage sources. In the case of the former, understanding the impacts of high flows in rivers and streams discharging to the recreational waters is crucial if accurate budgets are to be calculated and the impacts of expensive remedial measures are to be predicted at the scheme design stage.

If non-compliance has been historically associated with periods after rainfall events, then some attention to the remediation of catchment-derived sources of faecal indicators is prudent. Catchment modelling is one tool which can be used for the assessment of non-point source pollution and this approach is under development with the aim of generic model building to guide new scheme design.

REFERENCES

Anon. 1976. Council Directive 76/160/EEC concerning the Quality of Bathing Water. *Official Journal of the European Communities*, **L31**(5.2.1976), 1–7.

Anon. 1985a. House of Commons Committee on Welsh Affairs. *Coastal Sewage Pollution. Minutes of Evidence*, 5th December 1984. HMSO, London.

Anon. 1985b. House of Commons Committee on Welsh Affairs. *Coastal Sewage Pollution. Minutes of Evidence*, 16th January 1985. HMSO, London. 117–167.

Anon. 1985c. House of Commons Committee on Welsh Affairs. *Coastal Sewage Pollution. Report and Proceedings I and II*, 12th December 1985. HMSO, London.

Anon. 1991. Council Directive of 91/271/EEC concerning Urban Waste Water Treatment. *Official Journal of the European Communities*, **L1**(4.1.90).

Anon. 1994a. Select Committee on the European Communities. *Bathing Water. House of Lords Session 1994–5, 1st Report with Evidence*, 6th December. HMSO, London.

Anon. 1994b. Proposal for a Council Directive concerning the Quality of Bathing Water. *Official Journal of the European Communities*, **C112**(22.4.94), 3–10.

Anon. 1995a. Select Committee on the European Communities. *Bathing Water Revisited. House of Lords Session 1994–5, 7th Report with Evidence*, 21st March. HMSO, London.

Anon. 1995b. Parliamentary Debates, House of Lords Official Report. *Hansard*, **564**(90), 18th May, 684–708.

Anon. 1995c. *Cost of compliance with proposed amendments to Directive 76/160/EEC*. Presented to the House of Lords Select Committee Enquiry, prepared for the DoE by Halcrow plc.

Cabelli, V.J., Dufour, A.P., McCabe, L.J. and Levin, M.A. 1982. Swimming associated gastro-enteritis and water quality. *American Journal of Epidemiology*, **115**, 606–616.

Fleisher, J.M. 1990a. Conducting recreational water quality surveys. Some problems and suggested remedies. *Marine Pollution Bulletin*, **21**(2), 562–567.

Fleisher, J.M. 1990b. The effects of measurement error on previously reported mathematical relationships between indicator organism density and swimming associated illness: a quantitative estimate of the resulting bias. *International Journal of Epidemiology*, **19**(4), 1100–1106.

Fleisher, J.M. 1991. A re-analysis of data supporting the US Federal bacteriological water quality criteria governing marine recreational waters. *Journal of the Water Pollution Control Federation*, **63**, 259–264.

Fleisher, J.M. 1992. US Federal bacteriological water quality standards: a re-analysis of the data on which they are based. In Kay, D. (Ed.), *Recreational Water Quality Management*, Volume I. *Coastal Bathing Waters*. Ellis Horwood, Chichester, 113–128.

Fleisher, J., Jones, F., Kay, D. and Morano, R. 1993. Setting recreational water quality criteria. In Kay, D. and Hanbury (Eds), *Recreational Water Quality Management*, Volume II. *Fresh Water*. Ellis Horwood, Chichester, 123–136.

Fleisher, J.M., Kay, D., Salmon, R.L., Jones, F., Wyer, M.D. and Godfree, A. F. 1996. Marine waters contaminated with domestic sewage, non-enteric illnesses associated with bather exposure in the United Kingdom. *American Journal of Public Health*, **86**(9), 1228–1234.

Kay, D. 1994. Summary of epidemiological evidence from the UK sea bathing research programme. In Select Committee on the European Communities, *Bathing Water. House of Lords Session 1994–5, 1st Report with evidence*, 6th December. HMSO, London, 32–37.

Kay, D. and Rees, G. 1997. Recreational water: review of trends and events. In Earl, (Ed). *Marine Environmental Management 1996*. Candle Cottage, Gloucestershire, 107–112.

Kay, D. and Wyer, M.D. 1997. Microbial indicators of recreational water quality. In Kay, D. and Fricker, C. (Eds), *Coliforms and E.coli: Problem or Solution*. Royal Society of Chemistry, 89–100.

Kay, D., Wyer, M.D., McDonald, A.T. and Woods, N. 1990. The application of water quality standards to United Kingdom bathing waters. *Journal of the Institution of Water and Environmental Management*, **4**(5), 436–441.

Kay, D., Fleisher, J.M., Salmon, R., Jones, F., Wyer, M.D., Godfree, A., Zelanauch-Jaquotte, Z. and Shore, R. 1994. Predicting the likelihood of gastro-enteritis from sea bathing: results from randomised exposure. *The Lancet*, **34**, 905–909.

Kay, D., Jones, F., Fleisher, J. *et al.* 1996. *Relevance of faecal streptococci as indicator of pollution*. Report to DG XI of the Commission of the European Communities. CREH, University of Leeds.

Kay, D., Wyer, M.D., Fleisher, J. and Fewtrell, L. 1997. Health Efects and Standards Design, scope for an *a priori* approach. In Maire, E.A., Lightfoot, N. and Ramaekers, D.A. (Eds), *Report of the Workshop held in Sitges, Spain*, 26–29 April 1997. European Commission Measurements and Testing Programme Report EUR 17801, 52–56.

Kay, D., Wyer, M.D., Fewtrell, L., Crowther, J., Jackson, G. and O'Neill, G. 1997a. Non-sewage sources of faecal indicator in near-shore bathing waters. *Water and Health*, **22**, 225.

Kay, D., Wyer, M.D., Fleisher, J., Fewtrell, L. and Godfree, A.F. 1999. Health effects from recreational water contact. *Water Science and Technology* (in Press).

McDonald, A.T., Kay, D. and Jenkins, A. 1982. Generation of faecal and total coliform surges in the absence of normal hydrometeorological stimuli. *Applied and Environmental Microbiology*, **44**(2), 292–300.

PHLS (Public Health Laboratory Service) 1959. Sewage contamination of coastal bathing waters in England and Wales: a bacteriological and epidemiological study. *Journal of Hygiene, Cambs.*, **57**(4), 435–472.

Stevenson, A.H. 1953. Studies of bathing water quality and health. *American Journal of Public Health*, **43**, 529–538.

USEPA (United States Environmental Protection Agency) 1976. *Quality Criteria for Water.* Washington DC.

USEPA (United States Environmental Protection Agency) 1986. *Ambient Water Quality Criteria for Bacteria – 1986.* EPA440/5-84-002, Office of Water Regulations and Standards Division, Washington DC.

WHO (World Health Organisation) 1972. *Health Criteria for the Quality of Recreational Waters with Special Reference to Coastal Waters and Beaches.* WHO, Copenhagen.

Wilkinson, J., Jenkins, A., Wyer, M.D. and Kay, D. 1995. Modelling faecal coliform dynamics in streams and rivers. *Water Research*, **29**(3), 847–855.

Wyer, M.D., Jackson, G.F., Kay, D., Yeo, J. and Dawson, H.M. 1994. An assessment of the impact of inland surface water input to the bacteriological quality of coastal waters. *Journal of the Institution of Water and Environmental Management*, **8**, 459–467.

Wyer, M.D., Kay, D., Jackson, G.F., Dawson, H.M., Yeo, J. and Tanguy, L. 1995, Indicator organism sources and coastal water quality: a catchment study on the Island of Jersey. *Journal of Applied Bacteriology*, **78**, 290–296.

Wyer, M.D., Kay, D., Dawson, H.M., Jackson, G.F., Jones, F., Yeo, J. and Whittle, J. 1996. Delivery of microbial indicator organisms to coastal waters from catchment sources. *Water Science and Technology*, **33**(2), 37–50.

Wyer, M., O'Neill, G., Goodwin, V., Kay, D., Jackson, G., Tanguay, L. and Briggs, J. 1997a. Non-sewage derived sources of faecal indicator organisms in coastal waters. In Kay, D. and Fricker, C. (Eds), *Coliforms and* E. coli: *Problem or Solution.* Royal Society of Chemistry, 120–132.

Wyer, M.D., Kay, D., Crowther, J., O'Neill, J.G. and Jackson, G. and Fewtrell, L. 1997b. Non-outfall sources of faecal indicator organisms affecting the compliance of coastal waters with Directive 76/160/EEC. *Water Science and Technology*, **35**(11), 151–156.

4

Beach Health Risk Assessment and Pollution Perception

C. Nelson, A.T. Williams, G. Rees, D.Botterill and A. Richards

INTRODUCTION

Sun, sea and sand are obviously an appealing cocktail for recreationalists. However, the general public are becoming more familiar with environmental issues, especially beach pollution. The south Wales coastline has had a high profile in the media over recent years with the grounding of the *Sea Empress* oil tanker in February 1996, when 70 000 tons of oil was spilled across 30 miles of Welsh coastline (Anon., 1996). Speculation over risks to health from bathing in UK coastal waters has been intense over the last few years (Kay *et al.*,1994), in accordance with American studies done in the 1980s (Cabelli, 1983).

Field work was undertaken at Whitmore Bay, a resort beach in south Wales, to investigate the health risk from bathing in sewage-contaminated waters. Beach water quality regularly fails to achieve mandatory standards laid down in the EC Directive concerning the quality of bathing waters (CEC, 1976). A prospective epidemiological study was employed establishing a control group of non-bathers (unexposed) to compare with a group of bathers who entered the water (exposed). In addition, the sample group was stratified to analyse potential non-water-related risk factors, such as age and gender. An assessment of water quality was undertaken by taking regular water samples throughout the survey days, testing for *E. coli* and faecal streptococci in accordance with European legislation. Bacteriophages were tested for at Welsh Water plc laboratories to cater for potential changes to current European legislation on coastal waters. In addition to objective scientific analysis of health and water quality,

Water Quality: Processes and Policy. Edited by Stephen T. Trudgill, Des E. Walling and Bruce W. Webb.
© 1999 John Wiley & Sons Ltd.

the study investigated the public perception to coastal pollution. Limited work has been carried out in this field, but most has concentrated on perception of river environments (for example House and Sangster 1991; Burrows and House, 1989). All data in the study were collected via a semi-structured questionnaire.

PHYSICAL BACKGROUND

The study presented in this paper was carried out on Whitmore Bay, the larger of two designated beaches situated in the port town of Barry, 10 miles west along the coast from Cardiff, in the Vale of Glamorgan (UK). The beach itself is mostly sand, south-facing on the Bristol Channel, covers a large area of 200 000 m^2, and is 800 m long and 250 m wide to low water. It is backed by a Victorian promenade separating the beach from a developed hinterland which includes a holiday camp, fairground attractions, mini-golf and amusement arcades. Holiday shops, cafes and toilets provide excellent facilities for tourists and day trippers. The Bay has a big catchment area for both locals, Barry Town having a population of 46 000 (VOG, 1996), and day trippers. Close proximity to the M4 provides ease of access from the Welsh valleys and Gwent region for day trippers and also for holiday makers staying at the resort.

METHODOLOGY

Perception of Coastal Pollution

A semi-structured questionnaire was utilised for obtaining data on the public's perception to coastal pollution. Issues addressed included both perception of shore-side debris, composed of general litter and sewage-related debris, and respondents were also asked their views on marine pollution. A checklist was detailed in the questionnaire listing a comprehensive set of both beach litter and offshore pollutants. Participants were requested to select from the list the three most offensive forms of pollution. Non-parametric chi-square (χ^2) statistical analysis was performed on this data at the $p = 0.05$ level.

Water Quality Analysis

Samples of water were taken at 2-hour intervals on the survey days during a window slot between 11.00am and 3.00pm, the period of highest swimmer density. The time span accounted for temporal changes in water quality across tidal fluctuations. To assess spatial variations, two points across the beach were sampled at points of highest swimmer density. Replicate water sampling using the membrane filtration technique was carried out at the University of Glamorgan microbiology laboratory. The water was tested for both *E. coli* and faecal streptococci in line with the current EC Directive concerning the quality of bathing waters (CEC, 1976). This directive is currently under reform (CEC, 1994), with potential in the future to include F-specific RNA phages as an indicator of sewage contamination of coastal waters. To accommodate this possible

Table 4.1 *Exposed vs unexposed 2 × 2 cross-tabulation*

	Ill	Not ill
Exposed	a	b
Unexposed	c	d

$$\psi = \frac{ad}{bc}$$

addition to the bathing water directive, water samples were also tested for F-specific RNA phages by Welsh Water plc at Acer Laboratories, Bridgend.

Health Risk Assessment

Health risk from swimming at Whitmore Bay was calculated using odds ratios (ψ). The odds ratio is a suitable measurement of risk assessment when dealing with a binary response variable (Collett, 1991). By definition, the odds ratio is a ratio of the probability of contracting an illness (p_1) over the probability of not contracting an illness ($1 - p_1$) for the exposed group, over the probability of contracting an illness (p_2) over the probability of not contracting an illness ($1 - p_2$) in the unexposed group, denoted by:

$$\psi = \frac{p_1(1 - p_1)}{p_2/(1 - p_2)}$$

The exposed group are those respondents in the survey that entered the water and the unexposed are those that refrained from entering. The odds ratio for the exposed and unexposed groups can be calculated from a 2 × 2 cross-tabulation shown in Table 4.1 (Schlesselman, 1992).

The approximate 95% confidence limits indicate the reliability of the odds ratio, given by:

$$\psi_{\text{lower limit}} = \psi \exp [- 1.96 \sqrt{\text{var} (\ln \psi)}]$$
$$\psi_{\text{upper limit}} = \psi \exp [+ 1.96 \sqrt{\text{var} (\ln \psi)}]$$
$$[\text{var} (\ln \psi) \sim (1/a + 1/b + 1/c + 1/d)]$$

RESULTS AND DISCUSSION

Health Risk Assessment

A crude estimation of the extent to which exposure had an effect on illness is computed from Table 4.2. The odds ratio shows a significant elevation in symptom rates (21.2) among those that entered the water compared to the control group of those that refrained from entering. However, this result takes no account of risk variables,

Table 4.2   *Exposure and illness*

	Ill	Not ill
Exposed	94	291
Unexposed	3	197

$\psi = 21.21\ (6.65, 67.6)$

outlined in the following sections. A relative comparison of illness probabilities is achieved through stratification by age, gender and visitor type. Rates of illness amongst different socio-economic classes are also investigated, but only in terms of risk for the exposed group. Zero entries in the unexposed group result in the generation of infinite odds ratios between exposed and unexposed groups with regard to illness.

Stratified by Age

Initially respondents were required to indicate their age on the questionnaire by ticking the appropriate box. Ages were grouped in 10 year ranges. To avoid zero entries, which occurred in some of the unexposed groups, the six age group categories were collapsed down, forming three sets, 0–19, 20–39 and 40 + .

It is apparent that the older age category are least at risk from swimming, followed by the 0–19 age category. The most susceptible group of people to illness is the mid-age group 20–39 (Table 4.3).

Stratified by Gender

The results showed no statistical difference suggesting that neither gender group has a higher or lower risk from swimming in the sea (Table 4.4).

Table 4.3   *Illness in relation to age group (in years)*

	0–19		20–39		40 +	
	Ill	Not ill	Ill	Not ill	Ill	Not ill
Exposed	44	126	22	85	11	76
Unexposed	1	48	1	89	1	64

$\psi_{0-19} = 20.95\ (2.81, 150.33)$
$\psi_{20-39} = 27.22\ (3.61, 205.06)$
$\psi_{40-60+} = 10.95\ (1.39, 86.0)$

Table 4.4 *Illness in relation to gender*

	Male		Female	
	Ill	Not ill	Ill	Not ill
Exposed	37	90	46	197
Unexposed	1	46	2	155

ψ_{male} = 19.9 (2.65, 149.49)
ψ_{female} = 21.6 (5.19, 89.95)

Table 4.5 *Illness in relation to visitor type*

	Day tripper		Local	
	Ill	Not ill	Ill	Not ill
Exposed	76	232	18	52
Unexposed	2	148	1	46

$\psi_{day\ tripper}$ = 22.33 (5.39, 92.44)
ψ_{local} = 23.13 (3.02, 177.12)

Stratified by Visitor Type

In the context of visitor type, the day tripper represents beach users who travelled over 10 miles to reach their destination as opposed to locals who lived in closer proximity to the beach (Table 4.5). No significant difference was apparent between the two groups.

Stratified by Socio-Economic Status

As mentioned, only a comparison of risk can be viewed for the exposed group, stratified by socio-economic class (Table 4.6). Risk measures the association of illness occurring as a proportion of the population at risk. Students, which include children, had a significantly higher risk of contracting an illness, especially compared to the

Table 4.6 *Risk of illness in relation to socio-economic group*

	Employed		House wife		Student		Unemployed and retired	
	Ill	Not ill	Ill	Not ill	Ill	Not ill	Ill	Not ill
Exposed	31	109	35	105	22	36	6	34
Risk		22.14		25.00		37.93		15.00

unemployed and retired category. There was not a particularly large difference between employed people and housewives.

Water Quality Analysis

Figure 4.1 shows the geometric mean levels of *E. coli* and faecal streptococci counts
over the survey days during August. Counts for both *E. coli* and faecal streptococci
averaged 3400/100 ml and 440/100 ml, respectively (Nelson *et al.*, in press). Maximum
counts for the coliform bacteria reached a massive 45 000/100 ml and the faecal
streptococci count reached 20 300/100 ml. These figures far exceed the criteria set by
the EC bathing water directive, the mandatory level for *E. coli* being 2000/100 ml with
a guideline of 100/100 ml and the guideline standard for faecal streptococci 100/100 ml.
Zero colonies of F-specific RNA phage were found, which might be due to analytical
procedures. More alarmingly, in a study produced by Kay *et al.* (1994), a level of faecal
streptococci in excess of 32/100 ml was deemed to show a significant increase in
reported incidence of illness. Bacterial counts varied greatly over time, which might
well be due to a very high range of the tide in the Severn Estuary (University of Wales,
1997). There was also a trend of higher bacterial counts following the weekend.
Although these observations were only over a two week period, a possible reason may
be high visitor loads at the weekend.

Perception Analysis

There were difficulties in assessing people's perception to shore-side debris on the
actual beach because it was raked each morning. By the end of the day there were
significant volumes of litter left by visitors. The most prominent debris items noted on
the beach by respondents were: food packaging (82.3), plastic bottles (68.4), aluminium
cans (65.2), excrement (27.4) and hygiene items (22.7), the numbers in parentheses
representing the percentage score for each item.

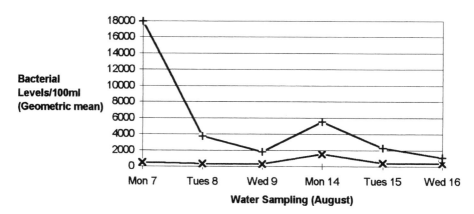

Figure 4.1   *Temporal bacterial trends: +, E. coli; ×, faecal streptococci*

Composition of general litter and sewage-related debris was found to be the most offensive concoction, in contrast to individual generic classes. The most sensitive groups of people to beach litter were females, the age range 30–39 years, and locals compared to visitors who travelled more than 10 miles to their destination (Williams and Nelson, 1997).

A high degree of worry was expressed by the public over the water condition, such that a large number (69%) actually decided not to enter, believing it to be polluted. Chi-square analysis at the 0.05 level showed females and also the age category 30–39 year to be more sensitive to perception of pollution. However, parents still chose to visit the beach for the sand and amenity value without allowing their children in the water. Water quality was the main reason for not swimming (54.9%), as opposed to temperature (23.3%), which was next. Floating objects were identified as being the most obtrusive forms of marine debris, being selected by 53% of the respondents, encompassing anything from food packaging and hygiene items to faecal matter. Colour of the water was reported to be bad by 21% of those surveyed. This is very likely due to the high sediment load in the water, being within the boundary of the Severn Estuary (University of Wales, 1997). Next in line was foul smells (14%) and then oil (12%).

CONCLUSION

Aesthetic value of land and seascapes are very important to recreationalists, and must be acknowledged. In particular, litter and sea pollution are obviously of great concern to those that use the beach. Increased public awareness of the environment could well deter people from visiting the coastline; this has great potential to harm the tourism industry, which would be disastrous in Wales. It is imperative that future beach management plans consider the users' perception of the coastline. Although poor visual appearance of the beach does not necessarily infer danger to health, results of the water quality aspects from this study strongly suggest a notable increase in incidence rates of illness from immersion in seawater at Whitmore Bay. These findings are in congruence with work done by Kay *et al.* (1994). His findings showed that 32/100 ml of faecal streptococci significantly increased the risk of gastrointestinal symptoms from swimming; this level is over 10 times lower than the average value of faecal streptococci found in this study.

The field work was carried out on a single beach, so the findings are obviously site specific. However, the fact that these illness rates are similar to evidence provided by other studies (Jones *et al.*, 1993; Cabelli *et al.*, 1982; Cabelli, 1983), makes it imperative that health risks to beach users are seriously addressed and appropriate regulatory criteria formulated. It will be interesting to see how potential reforms to the current EC bathing water directive effect change. On a regional scale, cleaner water quality is clearly on the horizon in Wales with Welsh Water's strategy to introduce new technology sewerage systems around the whole of the Principality (Lowe, pers. comm.). The Vale of Glamorgan Borough Council are very proactive in making sure the beach at Whitmore Bay is raked each morning, but improvements to water quality are not likely to be realised until at least the turn of the century.

REFERENCES

Anon. 1996. Sea Empress. *Inshore*, **3** (Summer).

Burrows, A.M. and House, M.A. 1989. Publics perception of water quality and the use of water for recreation. In Laikari, H. (Ed.), *River Basin Management*. Pergamon, Oxford. 371–379.

Cabelli, V.J. 1983. *Health effects criteria for marine recreational waters*. EPA 600/1-80-031, US Environmental Protection Agency, Health Effects Research Laboratory, North Carolina.

Cabelli, V.J., Dufour, A.P., McCabe, L.J. and Levin, M.A. 1982. Swimming-associated gastroenteritus and water quality. *American Journal of Epidemiology*, **115**(4), 606–616.

CEC. 1976. Council Directive of 8 December 1975 concerning the quality of bathing water (76/160/EEC). *Official Journal of the European Communities*, **L31/1** (5 February 1976).

CEC. 1994. Proposal for a Council Directive concerning the quality of bathing water. Document COM(94) 36 final.

Collett, D. 1991. *Modelling Binary Data*. Chapman and Hall, London.

House, M.A. and Sangster, E.K. 1991. Public perception of river-corridor management. *Journal of the Institute of Water and Environmental Management*, **15**(3), 312–317.

Jones, F., Kay, D., Wyer, M.D., Fleisher, J., Salmon, R. and Godfree, A.F. 1993. *Final report of the controlled cohort investigations into the health effects of bathing in sewage contaminated coastal waters*. Water Research Centre and Centre for Research into Environment and Health, University of Wales, Lampeter.

Kay, D., Fleisher, J.M., Salmon, R.L., Jones, F., Wyer, M.D., Godfree, A.F., Zelenauch-Jacquotte, Z. and Shore, R. 1994. Bathing water quality. *The Lancet*, **344**, 905–909.

Nelson. C., Williams, A.T. and Botterill, D. (in press). Quality of bathing waters in the coastal zone: A European and regional perspective.

Schlesselman, J.J. 1982. *Case-Control Studies: Design, Conduct, Analysis*. Oxford University Press, New York.

University of Wales. 1997. *Severn Estuary. Joint Issues Report*. 177pp.

VOG. 1996. *Vale of Glamorgan Fact Sheet*. Vale of Glamorgan Borough Council.

Williams, A.T. and Nelson, C. 1997. The public perception of beach debris. *Shore and Beach*, (July), 17–20.

5

The Validation of Pesticide Leaching Models for Regulatory Purposes

Adrian Armstrong, Tim Jarvis, Graham Harris and John Catt

INTRODUCTION

Pesticides are an important component of modern agriculture, being required to protect plants from the unwelcome attentions of weeds, pests and diseases which would otherwise result in significant reductions in crop yield or quality. The use of plant protection products has thus often been seen as a major plank in the continued struggle to feed an ever-expanding world population. However, pesticides are also, by definition, lethal products, and they can have effects beyond their immediate target. Ideally pesticides should be applied to the target only, and should not move any further. Inevitably, however, a small proportion may find its way to non-target ecosystems, and cause damage. Thus pesticides may affect non-target organisms at the point of application, or they may migrate away from that place of application into other areas, and so affect non-target ecosystems. This paper is concerned with the second risk, that of movement of pesticides to non-target systems.

Because of the potential damage to non-target systems, and in particular because of the risk to human health, the use of pesticides is strictly regulated. A major concern is that pesticides might find their way into potable water sources (e.g. Fielding *et al.*, 1991), and as a result, the maximum level for an individual pesticide in drinking water is specified as $0.2 \mu g/l$ by the EC directive on Drinking Water Quality (Anon., 1980).

Water Quality: Processes and Policy. Edited by Stephen T. Trudgill, Des E. Walling and Bruce W. Webb.
© 1999 John Wiley & Sons Ltd.

Throughout Europe, authorisation for the use of a particular pesticide needs to be granted by the relevant authority in accordance with EC directive 91/414 (Anon., 1991; Klein *et al.*, 1993). With respect to agricultural pesticides in the UK, the Pesticides Safety Directorate acts on behalf of five government departments. The process of registration requires the suppliers of the pesticide to present to the regulators evidence that the pesticide is safe to use, presenting no major risk to non-target ecosystems (neither human health nor other ecosystems) when applied according to label instructions. A major component in evaluating the safety of a pesticide is thus the consideration of its mobility in the environment. There are a number of major routes for the migration of pesticides away from the area of application. These include:

- spray drift on application
- gaseous losses to the atmosphere
- surface runoff, either in solution or bound on soil particles
- leaching to deep groundwater or drainage to surface waters.

The last of these, the leaching loss, is perhaps the most difficult to evaluate, and so is frequently the subject of modelling studies. This paper considers the use of models to predict the leaching of pesticides, particularly within the context of the registration process. Models for the prediction of pesticide movement in erosion, runoff and surface water are thus not considered explicitly in this paper, although the same general principles might be expected to apply to them.

As part of the registration process the applicant is required to submit to the regulatory authority data which address the environmental fate and behaviour of the pesticide. The dossier submitted usually includes laboratory and field studies and sometimes lysimeter and modelling studies. Models may be used at several stages in the development, testing and registration of pesticide (Hutson, 1992; Klein, 1991a). For example, simple models may also be used at the development stage of a compound as a screening tool in order to consider possible environmental problems. However, this paper is concerned with the use of models to simulate the environmental fate and behaviour of pesticides. In fact, most modelling studies submitted in support of registration concentrate on the likelihood of pesticide leaching to receiving waters (Jarvis, 1995). Such studies provide confidence that an unacceptable effect on the environment, including the pollution of water, will not occur. Since modelling studies have the advantage of being considerably cheaper to undertake than complex lysimeter or field leaching studies, they are often used by companies as a method of addressing possible leaching concerns for a particular compound. In addition, they have the second advantage of addressing simplified, generalised situations without the complications of soil and climatic variability.

When modelling studies of leaching are submitted as part of the regulatory submission, much of the input data are derived from laboratory experiments, due to the detailed information available from that source. Also, the pesticide product has a specified proposed use that will be expected, by most authorities, to be taken into account in the production of a realistic worst-case scenario for the simulation. The intention in using pesticide-specific and environmental inputs is that the outputs are as close to reality as possible, and this is essential since the results may be used in a quantitative manner.

In this context, several regulatory agencies have suggested that models can be used to indicate the safety of pesticides with respect to two major routes of movement: the deep leaching to groundwater, and the subsurface lateral movement via drainage systems to surface waters. Both these processes require the use of leaching models, which predict the concentration of pesticides in water draining either out of the base of a soil profile or laterally through the drainage system. Typical among these is the United States Environmental Protection Agency model, PRZM (Carsel *et al.*, 1984). Other states in Europe have adopted other models as standard, e.g. PELMO (Klein, 1991b, 1994) in Germany and PESTLA (Brouwer, 1994) in The Netherlands, and in particular have also identified standardised scenarios against which all substances will be tested. The problem with such scenarios is that they are difficult to define, and may not be appropriate for some circumstances. For example the German experience is built around the assumption that the maximum pesticide leaching risk is to ground-water by leaching through sandy soils, and takes no account of the phenomena of by-pass flow and the pollution of surface waters. Nevertheless, standardised scenarios are becoming an increasingly common component of the registration scene, and it is important that they are well defined and their relevance clearly identified. The combi-nation of a model and a soil scenario recognises the fact that different models perform with different degrees of success in different field situations (see for example the comparisons of Bergström and Jarvis (1994) and Walker *et al.* (1995)) and indeed between modellers (Brown *et al.*, 1996).

However, in all modelling studies, whether using standard models and standard scenarios, or new models and non-standard conditions, the regulatory agencies need to be able to evaluate the results of the modelling studies passed to them. They thus need to know the details of the modelling procedures on which the results submitted to them are based, the validity of the models, and the extent to which the results can be expected to relate to most situations. Until the models are better evaluated, and more confidence in their results is built up, models can only be used as an adjunct to field data, complement-ing and extending the results of the necessarily limited field results (Jarvis, 1995). Indeed it is hard to see that models will ever be a complete replacement for field studies.

Model validation is an unfortunately ambiguous term, and can be used in a number of different ways. This is because modelling is itself a problem-solving tool, and can be used to solve different classes of problems. We can perhaps identify three major uses of pesticide leaching models, which might be termed:

- Scientific modelling – to advance scientific understanding, testing of hypotheses, and storing knowledge of behaviour
- Management – to test the effects of treatments
- Regulatory – to examine the possibility of pesticides leaching into the environ-ment, with possible implications for the receiving ecosystems, and for the quality for drinking water.

There is thus a contrast between scientific validation, which is concerned to establish whether a model advances the understanding of the processes and the behaviour of pesticides in soils, and regulatory use, which is concerned to know the safety of that pesticide in the environment. However, it is generally accepted that detailed models based on physical principles are potentially useful for indicating the likely risk of pesticide movement.

SCIENTIFIC VALIDATION

The process of model validation should be distinguished from that of verification, which is concerned with the correct function of the algorithms and coding within the program (Fishman and Kiviat, 1968; Mihram, 1972). Validation is a matter of acquiring confidence that the model is an adequate representation of the real situation. The process of model validation involves placing the model results beside observations, and then judging whether the model represents these observations adequately. Model validation studies thus require large amounts of data, both input data to describe the system being modelled, and output data to compare with the model outputs.

An example has been the testing of the newly developed CRACK-NP model (Armstrong *et al.*, 1995a, 1996c) against the data from the Brimstone Farm experimental site (Harris *et al.*, 1994). Because the soil of the site is a well structured pelo-stagnogley (Avery, 1980), which typically has 55–60% clay and extensive macropores, standard solute movement models do not perform well at the site, and so the CRACK model (Jarvis and Leeds-Harrison, 1987; Jarvis, 1989) was used as a basis for modelling pesticide movement. CRACK conceptualises the soil as layers of aggregates. Within each layer, the soil porosity is divided into two components, macro- and micropores. Water flow into the macropores is generated at the surface by an infiltration excess and, once in the macropores, water moves rapidly downwards. The model allows for drainage of water out of the macropores, and so can model the mole-drainage system installed at Brimstone Farm. (Such mole-drainage systems are commonly used to drain fields on clay soils in the UK.) In CRACK, the macropores are defined by the inter-ped boundaries, and the movement of water and solute into the micropores is described by infiltration theory; water moves into the peds by sorption only and is extracted only by the crop. The important term describing the rate of interaction between the macropores and the peds is derived from observations of ped size, or its inverse, crack spacing. A pesticide module was added to the CRACK model, based on the descriptions contained in the CALF model of Nicholls *et al.* (1982), as modified by Walker (1987). Degradation is modelled using first-order kinetics, with coefficients dependent on temperature and moisture content, requiring the estimation of soil temperatures from air temperatures using the techniques of Walker and Barnes (1981). The combined model (with the addition of nitrate leaching routines) is now called CRACK-NP (Armstrong *et al.*, 1996a). CRACK-NP is intended for application only to soils which are dominated by preferential flows through macropores.

Because of the large analytical effort to determine the pesticide concentrations in both soil and water samples, the pesticide data are inevitably less frequent than either observations of other variables or predictions from the model. Table 5.1 illustrates this problem by listing the number of observations and model predictions for a single

Table 5.1   *Numbers of observational data points and model prediction points for a 120-day period for one plot at the Brimstone Farm site*

	Number of days with observations	Data points per day	Number of data points	Number of model predictions
Water discharge	120	48	5760	5760
Water table	120	24	2880	5760
Nitrate concentration	120	8	960	5760
Soil moisture contents	6	10	60	17 280
Pesticide in drain flow			20–30	5760
Pesticide in soil	6	6	36	17 280

plot of the Brimstone Farm site for a 120-day period. In this example, drain flows are recorded every half hour, water tables every hour, and nitrate concentrations every 3 hours, but pesticide concentrations are available for no more than 20 to 30 times per year. By contrast, the CRACK-NP model is capable of predicting the pesticide concentrations in the drainage water and in every soil layer at every model increment (normally every half hour). A major problem with this and many other pesticide leaching modelling studies was thus the scarcity of observations available for model validation. The problem is made worse by the fact that the relatively sparse pesticide data are also the most important. It was thus necessary to devise a validation strategy for a situation in which the number of model data points roughly matches the number of observations for some components of the model (specifically the hydrology) but is sparse for other components. Consequently we advocated a multistage cumulative validation procedure (Armstrong *et al.*, 1996a,b), commencing with hydrological validation, to take advantage of the more frequent hydrological data to test components of the model independently.

A standard statistical approach to model validation might have been to develop some global function of the differences between observed and predicted behaviour, to define a "goodness of fit" statistic (GOF), such as the sum of the squared differences (the error). The model can then be "fitted" to the data by altering the parameters in such a way as to minimise the error. However, such "fitted" models have parameters that are not independent of the "training" data set from which they were developed, and are thus difficult to apply outside that training data set.

Where the number of data points varies between variables, such "fitting procedures" are biased towards the data series with most data points. Consequently, if the model were to be optimised against the data described in Table 5.1, the resultant goodness of fit statistic would reflect the hydrological component of the model almost exclusively, and the difference between good and poor fits to the pesticide fate model would be virtually undetectable.

The solution to the problem of model validation that we have adopted is termed "multistage validation" (Armstrong *et al.*, 1996a,b), in which the model is validated one component at a time, and the validation confidence carried forward. The issue of the number of data points available for each validation stage is thus addressed separately.

Stage 1: Parameterisation of the Model using Independent Measurements

From the outset, any model validation exercise should be based on independent physical parameters. This places significant requirements on the data set used to verify the model, and also requires that the model parameters are capable of independent measurement. For model validation, the choice of a parameter set to represent the site must be taken *a priori*, and not retrospectively "tuned". Independently measured soil parameters from the experimental studies at Brimstone Farm are shown in Table 5.2. However, for other sites, such data are rarely available, and recourse might have to be made to "library" values of physical parameters, or the use of pedo-transfer functions to estimate them.

Stage 2: Hydrological Validation

Leaching is essentially a hydrological process, so any leaching model must reproduce the observed hydrology of the study site. There are generally two components of site hydrology: the soil water status and the fluxes of water through the profile (although these are of course related). In general, the model needs to be validated against both. If the fluxes simulated by the model are not correct, then estimates of pesticide movement will themselves be incorrect: the concentrations may be right but the total fluxes will be wrong, or *vice versa*. Only if the water fluxes are correct will both the concentrations and the solute fluxes be correctly estimated. The soil physical parameters from Table 5.2 were used to determine the soil water fluxes at Brimstone Farm. The results in Figure 5.1 show good simulation of drain flow events, both flow rate and timing. The less important differences between measured and simulated water tables at the first two days of the simulation are due to the limitations of the model in representing the water table below drain depth in dry periods. During the period of drain flow, the model predicts the general pattern of water table behaviour moderately well. Additional hydrological validation studies (Armstrong *et al.*, 1995a,b) showed that these input parameters reproduced the site hydrological behaviour (particularly the drain flow) well both over a whole winter and for short periods of time.

Table 5.2 *Some critical soil and pesticide (isoproturon) parameters for Brimstone Farm*

Layer	1	2	3	4	5	6	7
Depth (m)	0.05	0.10	0.20	0.30	0.40	0.60	1.00
Porosity (%)	55	55	55	55	55	50	50
Crack spacing (m)	0.05	0.10	0.10	0.20	0.20	0.30	0.40
Ped sorptivity at Wilting Point (mm/130 min)					12		
Hydraulic conductivity (mm/h)					40		
Soil bulk density (g/cm)					1.053		
Isoproturon K_d (mg/l)					2.9		
Isoproturon half-life: topsoil (days)					60		
Isoproturon half-life: subsoil, below 0.4 m (days)					200		

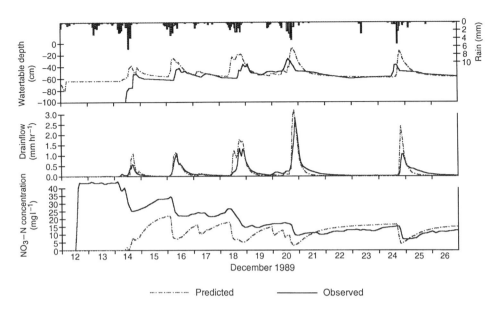

Figure 5.1   *CRACK-NP: predicted and observed nitrate concentrations, December 1989*

Stage 3: Solute Movement Validation

If at all possible, the solute movement component of the model should be independently tested in a similar way. Measurements of some ions can be made easily, quickly and cheaply, and so it is often possible to collect a sufficient density of data points to test the model rigorously. Where data permit, the model should thus be tested for solutes other than pesticide. The use of a conserved solute such as chloride or bromide over short periods is recommended. For the Brimstone Farm site it was possible to use the observations of nitrate concentrations in the drain water over a 10 day period in December 1989 to test the CRACK-NP model. Although nitrate is generated in the soil by mineralisation of organic matter, and some is lost by denitrification and plant uptake, when temperatures are low and biological processes slow, nitrate can be treated as a conserved solute over short periods. The model reproduced the pattern of nitrate leaching well, in particular modelling the diffusion-limited dilution of concentrations during peak drain flow events (Figure 5.1). Early disagreement in nitrate concentrations in Figure 5.1 are due to a lack of detailed knowledge of the distribution of nitrate within the profile at the start of the simulation period.

Stage 4: Fate of Pesticide in the Soil

All pesticide leaching models require some component to describe sorption of the pesticide onto the soil and its subsequent degradation by microbial activity, such as estimates of the K_d and half-life. However, estimation of these parameters is often far from straightforward. In particular the half-life needs to be estimated from incubation studies that relate to the properties of the site in question. For validation, it is important that models should use parameters obtained independently. The studies of Nicholls *et al.* (1993) for the Brimstone Farm site provided independent measurements of the crucial pesticide parameters included in Table 5.2 which were then used for the modelling exercise.

Stage 5: Pesticide Leaching Validation

Only in the last stage can the relatively small number of pesticide observations be placed against the simulations. Once it has been established that the model simulates the site hydrology and solute concentrations, then it is possible to evaluate the predicted pesticide losses. Because the pesticide observations are costly, the number of data points is usually small. Model simulations are generally acceptable if they are within an order of magnitude of the observations, and in particular if they match the observed patterns of behaviour. Figure 5.2 shows that the model simulated both the timing and peak concentrations of isoproturon in the drain water at Brimstone Farm.

The adoption of this five-stage strategy has permitted a validation of the CRACK-NP model for the Brimstone Farm site. Because the model uses independent parameters, and simulates both water and nitrate leaching behaviour, it is to be expected that the pesticide leaching predictions will be reasonably good, and this is borne out by the last comparison.

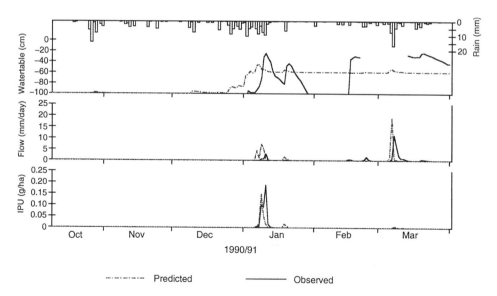

Figure 5.2   *CRACK-NP water table depths, drain flows and pesticide (isoproturon, IPU) loads versus observed data (after Armstrong et al., 1995a). Isoproturon was applied on 8 October 1990, and its subsequent fate and appearance in drainage water are reported by Harris et al. (1994)*

REGULATORY VALIDATION

Regulatory validation is concerned with a different set of criteria from those identified for the scientific validation. In addition to knowing that the model is valid for a specific pesticide at a specific site, it is also necessary to know that the model will make generalised predictions that will be valid for a number of sites and pesticides. The detailed single site, single data set validation, such as that illustrated for the CRACK-NP model at Brimstone Farm, has limitations. Although it is based on physical measurements and contains no "tuned" parameters, it is not known whether the model will apply to other locations. This level of validation is built up only by the repeated application of the model to multiple sites. At present, this multisite validation has not yet been performed for the CRACK-NP model.

However, the detailed considerations so far presented for the combination of the CRACK-NP model, the pesticide isoproturon and the Brimstone Farm site together present a scenario which can be considered validated. The issue of whether this scenario represents a common case for pesticide leaching risk is another question, which needs to be addressed in terms of the frequency of soils that have properties close to that of Brimstone Farm, and which have similar cultivation patterns. Although Brimstone Farm probably represents a "worst-case" scenario, in which the hydrology is dominated by macropore flow, the situation is not infrequent in the UK. Carter (pers. comm.) suggests that around one-third of all UK soils are cracking clays with a strong macropore component to their hydrology, and that many soils in addition have the potential for preferential movement. Data presented by Cannell *et al.* (1978) indicate that large areas similar to Brimstone Farm are suitable for the cultivation of cereals by the same methods, and maps of under-drainage (Robinson and Armstrong 1988) indicate that the drained condition at Brimstone Farm is also common for much of the same area. It is thus suggested that Brimstone Farm can be used as a type scenario for pesticide movement to surface waters via drainage.

The use of such validated scenarios has much to commend itself for pesticide fate studies, because the only variables that change are those relating to the pesticide itself, its physico-chemical properties and its time of use. As such it can perhaps be used without further calibration, although the results will of course be applicable to that scenario only.

The same model might then also be used to investigate the potential of management action on the rate and leaching of chemicals from the soil. Leeds-Harrison *et al.* (1992) and Matthews *et al.* (in press) used the CRACK-NP model to indicate that changing ped size can affect the total nitrate solute loads from clay soils, with smaller peds increasing the interaction between water and soil, and so increasing the total leaching amounts. Similar studies of the interaction between cultivation methods and the leaching of pesticides could also be undertaken at a theoretical level using the CRACK-NP model.

DISCUSSION

Models, no matter how good, are only simplifications and conceptualisations from the field situation. That is why we build them, because they are easier to manipulate and study than the field situation itself. Exact matching of the model to observation is probably never achievable. Indeed for regulatory purposes, such exact matching is not required. The regulatory use of a model is concerned with the generality of situations of which one field site is just an example. It is concerned with a single site only so far as it represents the greater generality.

For regulatory purposes models are required that can be used to estimate the likely fate and impact of new chemicals, new methods of application, or for new locations. Any concern over the approval of new pesticides needs to be related to their physical and chemical properties, their application rates and position in the cultivation cycle, and the management associated with them. The degree to which a model responds to the relevant input parameters indicates its suitability for testing of new substances; and the degree to which the model is capable of representing different management operations indicates its suitability for examining this option. The simulated impacts of pesticides at new untested locations depend on the degree to which the model represents the new situations, and the degree to which it is possible to parameterise them.

These needs are substantially different from the needs of scientific understanding, although confidence in a model is often derived from the credibility of its scientific basis. Models for registration require accuracy of prediction, rather than depth of understanding. However, the route to that accuracy is through scientific understanding, and for this reason, scientific development of pesticide fate models is required.

Lastly, environmental fate studies are concerned with issues of risk. There is thus a need to place the results from modelling studies within a stochastic conceptual framework. Probability modelling to identify the frequency of leaching risk from multiple model runs using distributions of input parameters has been described by Laskowski *et al.* (1990), but the computing requirements for the practical implementation of such a scheme are awesome. However, it is difficult to move directly from the paradigm of detailed mechanistic models which give single-value outputs from single-value inputs to a risk assessment without adopting such a scheme. Perhaps the alternative is to develop stochastic models directly, but this approach has not yet received acceptance.

ACKNOWLEDGEMENTS

The Brimstone Farm experiment is a joint venture of ADAS and IACR-Rothamsted. Funding from the Ministry of Agriculture, Fisheries and Food (MAFF) and British Agrochemicals Association is gratefully acknowledged. Modelling work has been funded by MAFF and the BBSRC/NERC Joint Initiative on Pollutant Transport in Rocks and Soils. We are grateful to Andy Matthews and Andrew Portwood of the ADAS Land Research Centre for their efforts in developing the model, to Professor Nick Jarvis of the Swedish University of Agricultural Sciences, Uppsala, for his help in making the code of CRACK available, and for his help and encouragement; to Peter Leeds-Harrison of the School of Agriculture, Food and Environment, Silsoe College, Cranfield University, and Tom Addiscott of Rothamsted Experimental Station for their help and advice in the development of the model. The errors remain our own. The opinions in this paper are, however, those of the authors and do not represent the official position of any organisation.

REFERENCES

Anon. 1980. EC Council directive relating to the quality of water intended for human consumption (80/778/EC). *Official Journal of the European Communities*, **L229**, 11–29.

Anon. 1991. Council directive of 15 July 1991 concerning the placing of plant protection products on the market (91/414/EEC). *Official Journal of the European Communities*, **L230**, vol. 34, 1–32.

Armstrong, A.C., Portwood, A.M., Harris, G.L., Catt, J.A., Howse, K.R., Leeds-Harrison, P.B. and Mason, D.J. 1995a. Mechanistic Modelling of pesticide leaching from cracking clay soils. In Walker, A., Allen, R., Bailey, S.W., Blair, A.M., Brown, C.D., Günther, P., Leake, C.R. and Nicholls, P.H. (Eds), *Pesticide Movement to Water*. BCPC Monograph No. 62, 181–186.

Armstrong, A.C., Addiscott, T.M. and Leeds-Harrison, P.B. 1995b. Methods for modelling solute movement in structured soils. In Trudgill, S.T. (Ed.), *Solute Modelling in Catchment Systems*. John Wiley, Chichester, 133–161.

Armstrong, A.C., Portwood, A.M., Leeds-Harrison, P.B., Harris, G.L. and Catt, J.A. 1996a. The validation of pesticide leaching models. *Pesticide Science*, **48**, 47–55.

Armstrong, A.C., Portwood, A.M., Harris, G.L., Leeds-Harrison, P.B. and Catt, J.A. 1996b. Multistage validation of pesticide leaching models In Del Re, A.A.M., Capri, E., Evans, S.P. and Trevisan, M. (Eds), *The Environmental Fate of Xenobiotics*, Proceedings of the X Symposium on Pesticide Chemistry, 30 September–2 October 1996, Castelnuovo Fogliani, Piacenza, Italy. La Goliardica Pavese, Pavia, Italy, 321–328.

Armstrong, A.C., Matthews, A.M., Portwood, A.M., Jarvis, N.J. and Leeds-Harrison, P.B. 1996c. *CRACK-NP: A model to predict the movement of water and solutes from cracking clay soils. Technical description and user's guide (Version 1.1)*. ADAS Land Research Centre, Gleadthorpe.

Avery, B.W. 1980. *Soil Classification for England and Wales (Higher Categories)*. Soil Survey Technical Monograph No. 14, Soil Survey of England and Wales, Harpenden.

Bergström, L. and Jarvis, N. (Eds) 1994. Special issue on the evaluation and comparison of pesticide leaching. *Journal of Environmental Science and Health, Part A – Environmental Science and Engineering*, **A29**(6).

Brouwer, W.W.M. 1994. Use of simulation models for registration purposes: Evaluation of pesticide leaching to groundwater in the Netherlands. *Journal of Environmental Science and Health, Part A – Environmental Science and Engineering*, **A29**(6), 1117–1132.

Brown, C.D., Baer, U., Günther, P., Trevisan, M. and Walker, A. 1996. Ring test with the models LEACHP, PRZM-2, and VARLEACH: variability between model users in prediction of pesticide leaching using a standard data set. *Pesticide Science*, **47**, 249–258.

Cannell, R.Q., Davies, D.B., Mackney, D. and Pidgeon, J.D. 1978. The suitability of soils for sequential direct drilling of combine-harvested crops in Britain: a provisional classification. *Outlook on Agriculture*, **9**, 306–316.

Carsel, R.F., Smith, C.N., Mulkey, L.A., Dean, J.D. and Jowise, P.P. 1984. *User's manual for the pesticide root zone model (PRZM) Release 1*. USEPA, EPA-600/3-84-109, US Government Printing Office, Washington, DC.

Fielding, M., Barcelo, D., Helweg, A., Galassi, S., Tortensson, L., Van Zoonen, P., Wolter, R. and Angeletti, G. 1991. *Pesticides in Ground and Drinking Water*. Water Pollution Research Report, CEC, DG XII, Brusssels.

Fishman, G.S. and Kiviat, P.J. 1968. The statistics of discrete event simulation. *Simulation*, **10**, 185–195.

Harris, G.L., Nicholls, P.H., Bailey, S.W., Howse, K.R. and Mason, D.J. 1994. Factors influencing the loss of pesticides in drainage from a cracking clay soil. *Journal of Hydrology,* **159,** 235–253.

Hutson, J.L. 1992. The use of models in the regulatory decision making process. *Brighton Crop Protection Conference – Pests and Diseases – 1992.* BCPC paper 10–2, 1253–1260.

Jarvis, N.J. 1989. *CRACK – a model of water and solute movement in cracking soils.* Department of Soil Sciences Report 159, Swedish University of Agricultural Sciences, Uppsala, Sweden.

Jarvis, N.J. and Leeds-Harrison, P.B. 1987. Modelling water movement in drained clay soil. I. Description of the model, sample output and sensitivity analysis. *Journal of Soil Science,* **38,** 487–498.

Jarvis, T.D. 1995. Simulation modelling and UK regulatory decision-making:– perfect partners or warring factions? *Pesticide Movement to Water* BCPC Monograph 62, 357–362.

Klein, A.W., Goedicke, J., Klein, W., Herrchen, M. and Kördel, W. 1993. Environmental assessment of pesticides under directive 91/414/EEC. *Chemosphere,* **26,** 979–1001.

Klein, M. 1991a. Application and validation of pesticide leaching models. *Pesticide Science,* **31,** 389–398.

Klein, M. 1991b. *PELMO: Pesticide Leaching Model.* Fraunehofer-Institut für Umweltchemie und Ökotoxikologie, D-5946 Schmallenberg, Germany.

Klein. M. 1994. Evaluation and comparison of pesticide leaching models for registration purposes. Results of simulations performed with the pesticide leaching model. *Journal of Environmental Science and Health, Part A – Environmental Science and Engineering,* **A29**(6), 1197–1210.

Laskowski, D.A., Tillotson, P.M., Fontaine, D.D. and Martin, E.J. 1990. Probability modelling. *Philosophical Transactions of the Royal Society of London, Series B,* **329,** 383–389.

Leeds-Harrison, P.B., Vivian, B.J. and Chamen, W.C.T. 1992. *Tillage effects in drained clay soils.* ASAE Paper 92–2648, Winter Meeting of the ASAE, 15–18 December 1992, Nashville.

Matthews, A.M., Portwood, A.M., Armstrong, A.C., Leeds-Harrison, P.B., Harris, G.L., Catt, J.A. and Addiscott, T.M. (in press). CRACK-NP, development of a model for predicting pollutant transport at the Brimstone Farm site, Oxfordshire, UK. *Soil and Tillage Research.*

Mihram, G.A. 1972. Some practical aspects of the verification and validation of simulation models. *Operational Research Quarterly,* **23,** 17–29.

Nicholls, P.H., Bromilow, R.H. and Addiscott, T.M. 1982. Measured and simulated behaviour of fluometuron, aldoxycarb and chloride ion in a fallow structured soil. *Pesticide Science,* **13,** 475–383.

Nicholls, P.H., Evans, A.A., Bromilow, R.H., Howse, K.R., Harris, G.L., Rose, S.C., Pepper, T.J. and Mason, D.J. 1993. Persistence and leaching of isoproturon and mecoprop in the Brimstone Farm plots. *Weeds: Proceedings of the Brighton Crop Protection Conference 1993.* British Crop Protection Council, 849–854.

Robinson, M. and Armstrong, A.C. 1988. The extent of agricultural field drainage in England and Wales, 1971–80. *Transactions, Institute of British Geographers, New Series,* **13,** 19–28.

Walker, A. 1987. Evaluation of a simulation model for prediction of herbicide movement and persistence in soil. *Weed Research,* **27,** 142–152.

Walker, A. and Barnes, A. 1981. Simulation of herbicide persistence in soil: a revised computer model. *Pesticide Science,* **12,** 123–132.

Walker, A., Calvet, R., Del Re, A.A.A.M., Pestemer, W. and Hollis, J.M. 1995. *Evaluation and improvement of mathematical models of pesticide mobility in soils and assessment of their potential to predict contamination of water systems.* Mitteilungen uas der Biologischen Bundesanstalt für Land- und Forstwirtschaft, Heft 307, Berlin-Dahlem.

6

Water Quality Processes in Catchments: An Integrated Modelling Approach for Scenario Analysis

P.G. Whitehead, E. Wilson and D. Butterfield

INTRODUCTION

In the recent Hydrological Society Penman Lecture, Professor Uri Shamir highlighted the role of water quality and hydrology in resolving local, national and international conflicts: quantitative research tools can assist in understanding the complex interactions and dynamic behaviour and can then be used in scenario analysis to assess alternative management strategies (Shamir, 1997). Models such as MAGIC (Model of Acidification of Groundwaters in Catchments; Cosby *et al.*, 1985a,b) and QUASAR (Quality Simulation Along Rivers; Whitehead *et al.*, 1997) have been used extensively to resolve scientific issues concerning acidification and river pollution, respectively. Moreover these models have been extensively used to assess management options and to resolve conflicts of interest.

In this paper a new model is described which addresses the issues of atmospheric nitrogen deposition, agricultural fertiliser application and effluent discharges in river basins. The overall balance of nitrogen in a catchment will vary according to these three major sources as well as the availability of nitrogen from natural biochemical processes such as mineralisation and nitrification. The new model INCA (Integrated Nitrogen in Catchments) seeks to account for all the nitrogen sources, sinks and processes in

Water Quality: Processes and Policy. Edited by Stephen T. Trudgill, Des E. Walling and Bruce W. Webb.
© 1999 John Wiley & Sons Ltd.

catchments. The model is described in detail and illustrated by way of application to the River Tywi in south Wales. Scenario analysis is used to demonstrate the applicability of the model to evaluate the impacts of changed land use and nitrogen deposition.

NITROGEN MODELLING

There has been considerable research on nitrogen processes and models in recent years (DOE, 1994; Skeffington and Wilson, 1988). However, most of these models have been directed at a specific problem or process. For example there are excellent models of upland systems such as MAGIC (Cosby *et al.*, 1985ab), MAGIC-WAND (Jenkins *et al.*, 1997) and MERLIN (Cosby *et al.*, 1997) and models developed for lowland agricultural systems (Addiscott and Whitmore, 1987; Cooper *et al.*, 1993).

In the case of the upland models these address the issue of SO_x and NO_x deposition and impacts on upland soils, streams and lakes. They are long-term, process-based models and have been used for assessing emission scenarios, critical loads and impacts of land use change. They do not, however, provide information on short-term variation in streams or the distributed behaviour across a catchment or down a river system.

The lowland agricultural models (Addiscott and Whitmore, 1987) are also process based but address the question of fertiliser impacts on soils, groundwaters and stream water. They tend to be dynamic models and can provide short-term (i.e. weekly) information. An additional nitrogen model for river systems exists that takes into account instream nitrification and denitrification processes and hence losses within rivers. One such model, QUASAR, has been developed by Whitehead *et al.* (1997) and applied to large river basins (Whitehead, 1990; Whitehead and Williams, 1984). A recent model by Lunn (1995) addresses the whole catchment in a distributed manner but is driven by a complex hydrological model, SHE (Systeme Hydrologique Europeen; Abbott *et al.*, 1986). In complete contrast the export coefficient method of Johnes (1996) uses an empirical approach to model nitrogen based on observations and historical records of land use. This had been a highly successful strategy for estimating annual or seasonal nitrogen loads. However, the model is not process based and cannot be used to generate daily nitrogen variations.

In order to achieve the objective of this study, it has been necessary to create a new generalised model so that atmospheric, upland, lowland and within-river processes together with hydrology can be addressed within a single integrated model. This model is:

- dynamic – so that daily time series of nitrogen behaviour are obtained and account taken of hydrological flowpaths and nitrogen processes;
- stochastic – so that parameter sensitivity can be assessed and outputs derived in probabilistic or percentile terms;
- semi-distributed – so that the spatial variations in land use, human impacts, effluent discharges and varying deposition levels can be taken into account.

The term semi-distributed is used here as it is not intended to model catchment land surface in a detailed manner. Rather, different land classes, subcatchment boundaries and land use areas will be modelled simultaneously and information fed sequentially into a multireach river model.

THE INCA MODEL

The INCA model has been designed to investigate the fate and distribution of nitrogen in the aquatic and terrestrial environment. The model simulates flow, nitrate-nitrogen and ammonium-nitrogen and tracks the flow paths operating in both the land phase and riverine phase. The model is dynamic in that the day-to-day variations in flow and nitrogen can be investigated following a change in input conditions such as atmospheric deposition/sewage discharges or fertiliser addition. The model can also be used to investigate a change in land use (e.g. moorland to forest or pasture to arable). Dilution, natural decay and biochemical transformation processes are included in the model as well as the interactions with plant biomass such as nitrogen uptake by vegetation.

INCA has been designed to be easy to use and fast, with excellent output graphics. The menu system allows the user to specify the semi-distributed nature of a river basin or catchment, to alter reach lengths, rate coefficients, land use, velocity–flow relationships and to vary input nitrogen deposition loads.

INCA provides the following outputs:

- daily time series of flows, nitrate-nitrogen and ammonium-nitrogen concentrations at selected sites along the river;
- profiles of flow or nitrogen along the river at selected times;
- cumulative frequency distributions of flow and nitrogen at selected sites;
- tables of statistics for all sites;
- daily and annual nitrogen loads for all land uses and all processes.

INCA runs on any IBM-compatible PC running MSDOS. Full details of the system requirements and installation are given elsewhere (Butterfield and Whitehead, 1997).

Figures 6.1 and 6.2 show the principal flow paths and processes operating; all processes are simulated using a mass balance differential equation approach. These equations are first order and describe the dominant processes operating together with process interactions. The equations are solved using a fourth order Runga Kutta method of solution with a Merson variable step length integration routine. This enables stable numerical integration of the equations and minimises numerical problems. The advantage of this scheme is that scientific effort can be directed to ensuring correct process formulation and interaction rather than numerical stability problems.

HYDROLOGICAL MODEL

The hydrological model consists of three parts. Firstly the MORECS soil moisture and evaporation accounting model (Meteorological Office, 1981) is used to convert daily

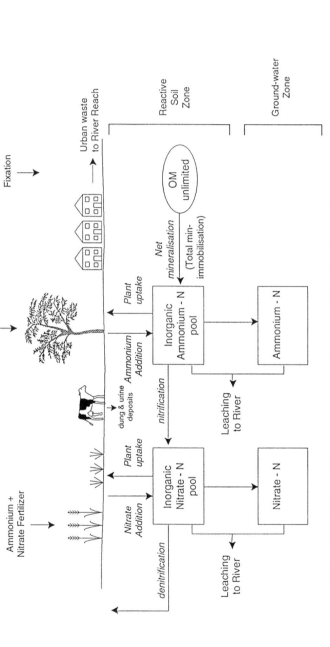

Figure 6.1 Nitrogen inputs, processes and outputs in the soil and groundwater system

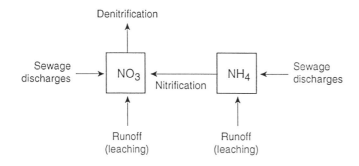

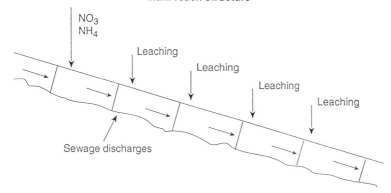

Figure 6.2   *Nitrogen inputs, processes and outputs in the river system*

rainfall data into an "effective" rainfall time series. By effective we mean the water that reaches the soil surface after allowing for interception and evapotranspiration losses. This hydrologically effective rainfall (HER) is used to drive the water transfers and nitrogen movement through the catchment system. The advantage of the MORECS model is that a daily time series of soil moisture deficit (SMD) is determined at the same time as HER and this information is essential for nitrogen process models as key processes such as mineralisations and denitrification depend on soil moisture levels.

The second component of the hydrological model is to simulate the land surface. There are many approaches to modelling land flow processes. These range from time series techniques (Whitehead *et al.*, 1979) through lumped hydrological models (Christophersen *et al.*, 1982) to fully distributed models such as SHE (Abbott *et al.*, 1986). In the current model development a semi-distributed approach has been adopted so that the dynamics of each subdrainage basin can be characterised and incorporated into the overall system model. Figure 6.1 shows the land cell model: it is necessary to know the principal response time of the reactive soil zone and the deeper groundwater zone as well as the flows moving through these zones. These are the two key reservoirs of water within the system, with complex nitrogen processes operating in the soil reactive zone and nitrogen being transported through the groundwater zone. The flow model for these two zones can be written as:

Soil zone:
$$\frac{dx_1}{dt} = \frac{1}{T_1}(U_1 - x_1) \tag{1}$$

Groundwater:
$$\frac{dx_2}{dt} = \frac{1}{T_2}(U_8 x_1 - x_2) \tag{2}$$

where x_1 and x_2 are output flows (m^3/s) for the two zones and U_1 is the input driving hydrologically effective rainfall. T_1 and T_2 are time constants (i.e. residence times) associated with the zones and U_8 is the baseflow index (i.e. proportion of water being transferred to the lower groundwater zone). The baseflow index information can be obtained from the hydrological year books (Institute of Hydrology, 1996) or from the application of time series modelling techniques.

Such a modelling technique, IHACRES, has been developed by Jakeman *et al.* (1990). IHACRES uses a recursive approach to time series analysis to determine the dynamics of the two separate hydrological zones, as illustrated in Figure 6.1, and can be calibrated against rainfall and runoff data for subcatchments to compute the baseflow separation, the soil zone time constant and the groundwater time constant.

The third component of the hydrological model is the river flow model. This model is based on mass balance of flow and uses a multireach description of the river system (Whitehead *et al.*, 1979, 1981, 1997). The flow variation in each reach is based on a non-linear reservoir model. The model may be viewed in hydrological flow routing terms as one in which the relationship between inflow, I, outflow, Q, and storage, S, in each reach is represented by:

$$\frac{dS(t)}{dt} = I(t) - Q(t) \tag{3}$$

where $S(t) = T(t)*Q(t)$; T is a travel time parameter, which can be expressed as:

$$T(t) = \frac{L}{v(T)} \tag{4}$$

where L is the reach length (m) and v, the mean flow velocity in the reach (m/s), is related to discharge, Q, through:

$$v(t) = aQ(t)^b \tag{5}$$

where a and b are constants to be estimated from tracer experiments or from theoretical considerations. Whitehead *et al.* (1986) review alternative methods of estimating velocity–flow relationships in rivers. They conclude that an empirical approach using tracer experiments is preferable to theoretical estimation techniques. This is because tracer experiments integrate the actual behaviour of a river channel whereas theoretical approaches are often limited by a need for physical channel information, which is often not available, and a detailed knowledge of channel roughness information. Estimation of Manning n, for example, is always rather subjective.

NITROGEN PROCESS EQUATIONS

The hydrological model provides information on the flow moving through the soil zone, the groundwater zone and the river system. Simultaneously, whilst solving the flow equations, it is necessary to solve the mass balance equations for both nitrate-nitrogen and ammonium-nitrogen in both the soil and groundwater zones. The key processes that require modelling in the soil zone, as indicated in Figure 6.1, are plant uptake for NH_4-N and NO_3-N, ammonia nitrification, denitrification of NO_3^-N, ammonia mineralisation, ammonia immobilisation and N fixation. All of these processes will vary from land use to land use and a generalised set of equations is required for which parameter sets can be derived for different land uses. The land phase model must also account for all the inputs affecting each land use including dry and wet deposition of NH_4-N and NO_3-N and fertiliser addition for both NH_4-N and NO_3-N (e.g. as ammonium nitrate). Also temperature and soil moisture will control certain processes so that, for example, nitrification reaction kinetics are temperature dependent and denitrification and mineralisation are both temperature and soil moisture dependent.

In the groundwater zone it is assumed that no biochemical reactions occur and that a mass balance of NH_4-N and NO_3-N is adequate. The equations used in INCA are as follows:

Nitrate-n:

Soil zone:
$$\frac{dx_3}{dt} = \frac{1}{T_1}(U_3 - x_1x_3) - C_3U_7x_3 + C_6x_5 - C_1U_5x_3 + C_2 \tag{6}$$

Groundwater:
$$\frac{dx_4}{dt} = \frac{1}{T_2}(x_3x_1U_8 - x_2x_4) \tag{7}$$

Ammonium-N

Soil zone:
$$\frac{dx_5}{dt} = \frac{1}{T_1}(U_4 = x_1x_5) - C_{10}U_7x_5 - C_6x_5 + C_7U_6 - C_8x_5 \tag{8}$$

Groundwater:
$$\frac{dx_6}{dt} = \frac{1}{T_2}(x_5x_1U_8 - x_2x_6) \tag{9}$$

where x_3 and x_4 are the daily NO_3-N concentrations (mg/1), in the soil zone and groundwater zone, respectively, and x_5 and x_6 are the daily NH_4-N concentrations (mg/1), in the soil zone and groundwater zone, respectively.

U_8 is the baseflow index and C_3, C_6, C_1, C_2, C_{10}, C_7, C_8 are rate coefficients (per day) for, respectively, plant uptake of nitrate, ammonia nitrification, nitrate denitrification, nitrate fixation, plant uptake of ammonia, ammonia mineralisation, ammonia immobilisation. U_3 and U_4 are the daily nitrate-nitrogen and ammonium-nitrogen loads entering the soil zone and constitute the additional dry and wet deposition and agricultural inputs (e.g. fertiliser addition). All rate coefficients are temperature dependent using the equation:

$$C_n = C_n 1.047^{(\theta_s - 20)}$$

where θ_s is soil temperature estimated from a seasonal relationship dependent on air temperature as follows:

$$\text{soil temperature} = \text{air temperature} + C_{16} \sin\left(\frac{3}{2}\pi \frac{day\ no.}{365}\right)$$

where C_{16} is the maximum temperature (°C) difference between summer and winter conditions (Green and Harding, 1980).

U_7 is a seasonal plant growth index (after Hall and Harding, 1993) where:

$$U_7 = 0.66 + 0.34 \sin\left(2\pi \frac{[day\ no. - C_{11}]}{365}\right)$$

where C_{11} is the day number associated with the start of the growing season.

U_5 is a soil moisture threshold below which denitrification will not occur. Denitrification generally will only be significant when soil moisture levels are high. Similarly U_6 is a soil moisture control for mineralisation which permits mineralisation when soil water content is above a threshold level.

Nitrogen process equation: river system

In the river the key processes are shown in Figure 6.2 to be denitrification of NO_3-N, nitrification of NH_4-N and mass balance. As shown in Figure 6.2 the reach mass balance must include the upstream NO_3-N and NH_4-N together with inputs from both the soil zone and groundwater zone as well as direct effluent discharges.

The equations for the flow NO_3-N and NH_4-N river reaches are then:

Flow:
$$\frac{dx_7}{dt} = \frac{1}{T_3}(U_9 - x_7) \tag{10}$$

Nitrate:
$$\frac{dx_8}{dt} = \frac{1}{T_3}(U_{10}U_9 - x_7 x_8) - C_{17}x_8 + C_{14}x_9 \tag{11}$$

Ammonia:
$$\frac{dx_9}{dt} = \frac{1}{T_3}(U_{11}U_9 - x_7 x_9) - C_{14}U_3 x_9 \tag{12}$$

where $T_3 = L/v = L/(aQ^b)$ is a non-linear velocity flow relationship as described previously; U_9 is the upstream flow (m³/s), U_{10} is the upstream NO_3-N (mg/l) and U_{11} is the upstream NH_4-N (mg/l). T_3 is the reach time constant (or residence time) which varies from day to day, x_7 is the estimated downstream flow rate (m³/s) and x_8 and x_9 are the downstream (i.e. reach output) concentrations of nitrate and ammonia, respectively, and C_{17} and C_{18} are temperature-dependent rate parameters for denitrification and nitrification, respectively. Temperature effects are introduced related to river water temperature, θ_w, as follows:

$$C_n = C_n 1.047^{(\theta_w - 20)}$$

Table 6.1 *Institute of Terrestrial Ecology*

Class	Land cover	Class	Land cover
1	Sea/Estuary	14	Shrub/Orchard
2	Inland Water	15	Deciduous Woodland
3	Beach	16	Coniferous Woodland
4	Salt Marsh	17	Upland Bog
5	Grass Heath	18	Tilled Land
6	Mown/Grazed Turf	19	Ruderal Weed
7	Meadow	20	Suburban Development
8	Rough/Marsh Grass	21	Continuous Urban
9	Moorland Grass	22	Inland Bare Ground
10	Open Shrub Moor	23	Felled Forest
11	Dense Shrub Moor	24	Lowland Bog
12	Bracken	25	Open Shrub Heath
13	Dense Shrub Heath		

APPLICATION TO RIVER TYWI SYSTEM

The River Tywi system in south Wales, as shown in Figure 6.3, drains southwest towards Carmarthen from the upland areas surrounding Llyn Brianne. The river system is 33 km in length and drains an area of 1090 km². The Llyn Brianne area has been the subject of a major acidification study (Edwards *et al.*, 1990) and extensive

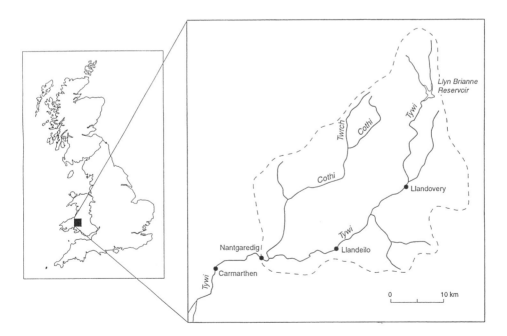

Figure 6.3 *Map showing River Tywi catchment*

Table 6.2   *Land use classification used in INCA*

Land use group no.	Land use group name	ITE land cover classes contained in group
1	Forest	14, 15, 16, 23
2	Short Vegetation–Ungrazed	8, 11, 13, 17, 19, 24
3	Short Vegetation–Grazed not Fertilised	5, 7, 9, 10, 12, 25
4	Short Vegetation–Fertilised	6
5	Arable	18
6	Urban	20, 21, 22

Table 6.3   *Reach and land use information used in INCA for the River Tywi*

Reach name	Reach length (m)	Area (km²)	Land use for INCA groups (%)*					
			1	2	3	4	5	6
1 Source	10	39	69	0	31	0	0	0
2 Tywi	5000	46	46	2	50	2	0	0
3 Brianne	3000	53	28	0	72	0	0	0
4 Doethie	1000	301	33	2	13	52	0	0
5 Galltybere	1250	20	55	5	40	0	0	0
6 Gwenffrwd	2500	126	55	1	15	28	0	1
7 Rhandirmyn	5000	2	50	0	50	0	0	0
8 Rhydwydd	2000	2	100	0	0	0	0	0
9 Gwenlas	3250	18	11	0	56	33	0	0
10 Dolachirian	3750	16	38	0	31	31	0	0
11 ABran	2500	63	29	0	10	61	0	0
12 Dolgarreg	6000	19	26	0	6	68	0	0
13 Sawdde	2250	98	37	0	9	54	0	0
14 Glanrhyd	5500	7	0	0	14	86	0	0
15 Pontbren	5000	138	33	2	22	42	0	1
16 Llandeilo	5750	63	33	0	42	23	2	0
17 Cilsen	6500	28	29	0	0	71	0	0
18 Dryslwyn	8500	18	11	0	0	83	6	0
19 Ffinnant	1250	12	42	0	0	58	0	0

*See Table 6.2 for land use classification (Nos. 1–6) used in INCA

modelling research has been undertaken on a range of different subcatchments. The focus of these studies was the impact of SO_x deposition, although extensive data sets were obtained for nitrogen and these data are of interest for the current application of INCA. The bedrock in the Brianne area consists of Ordovician and Silurian shales, grits and mudstones overlain by oligo-morphic peats and ferric stagnopodzols on the upper slopes and brown podzolic soils on the lower slopes. The land use in the Brianne area is either improved moorland or Spruce and Lodgepole Pine (*Pinas Contorta*). Further down the system the soils become brown earth and typical stagnogley soils and waters are significantly less acidic due to high pH waters draining calcite-rich subcatchments.

Nitrate and ammonium concentrations in catchments are affected greatly by land use. Nitrogen uptake, for instance, occurs at different rates in forested and grassland system. It makes sense, therefore, to use a different cell model parameter set for different land use classes. There are 25 separate land cover classes in the widely accepted Institute of Terrestrial Ecology (ITE) land classification system, so rather than use 25 different parameter sets, as shown in Table 6.1, these classes have been grouped together to form six principal groups, shown in Table 6.2, each containing land use classes of similar types.

FLOW AND NITROGEN SIMULATION RESULTS

The integrated hydrological model has been applied to the River Tywi system and Figure 6.4 shows the hydrologically effective rainfall (HER) calculated from the MORECS model, the air and soil temperature for the Tywi catchment and the soil moisture deficit (SMD). The dry summer months generate a major decline in soil moisture levels, particularly from day 130 through to day 221 when a major summer storm breaks the drought. The INCA model can generate daily simulated flows for any of the 19 reaches along the river system and model outputs for reach 19 for the River

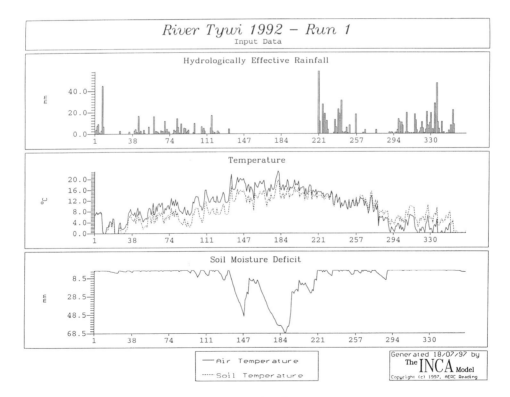

Figure 6.4 *Input data for River Tywi*

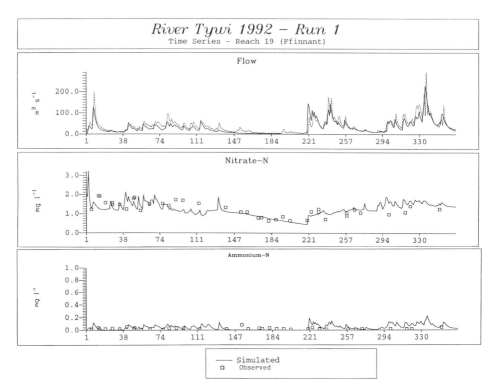

Figure 6.5   *Time series of simulated and observed flow, nitrate and ammonia for River Tywi, 1992*

Tywi are shown in Figure 6.5. The dynamics of flow dominate the hydrological and chemical response and Figure 6.5 indicates that the model is successfully simulating the flow.

Figure 6.5 indicates that there is considerable day-to-day variation in nitrogen flux depending on input distribution (e.g. fertiliser addition), hydrological flushing of nitrogen through the system and seasonal patterns of temperature and soil moisture deficit. For example denitrification only occurs when the soils are wet and nitrogen uptake only occurs during plant growth periods. Flushing of ammonia and nitrogen is controlled by the hydrological regime. Needless to say, these patterns will vary from year to year depending on rainfall, temperature, soil moisture and crop growth. The annual loads show that the fertiliser addition in the arable and surface vegetation is dominant but these loads will affect catchments differently depending on the distribution of land use.

As shown in Figure 6.5, nitrate and ammonia are simulated with reasonable accuracy especially the low nitrate and ammonia conditions in summer. At the end of the summer drought (on day 221, mid-August) the increased flows give rise to a pulse of nitrate according to the model. There have been sampling and modelling studies elsewhere that confirm this flushing effect in catchments (Whitehead and Williams, 1984; Whitehead, 1990).

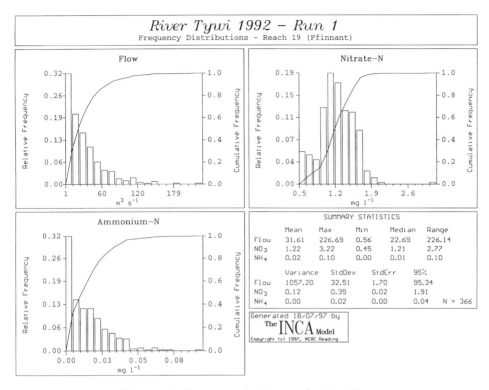

Figure 6.6 *Frequency distributions for River Tywi*

Figure 6.6 shows the distribution of flow, nitrate and ammonia for reach 19 on the Tywi and the model generates a changing distribution along the river system. The instream dispersion, dilution, denitrification and nitrification processes along the river alter the shape of the distribution and the nitrate concentrations vary considerably, reaching a peak in the middle reaches of the river adjacent to effluent discharges, and falling in the lower reaches as increased denitrification occurs in the lower sections of the river. The residence time is an important controlling parameter in the lower reaches allowing more time for denitrification as the water velocities fall. The reach-by-reach changes in concentration are calculated by the model, which will print out annual statistics for each reach.

SCENARIO ANALYSIS

INCA was designed to investigate alternative scenarios and two different scenarios have been investigated using INCA; these include changing land use and increasing deposition of nitrogen.

Figure 6.7 shows the land use change scenario. Reach 1 according to the land use classification has 69% forest and 31% short vegetation grazed, not fertilised. As a

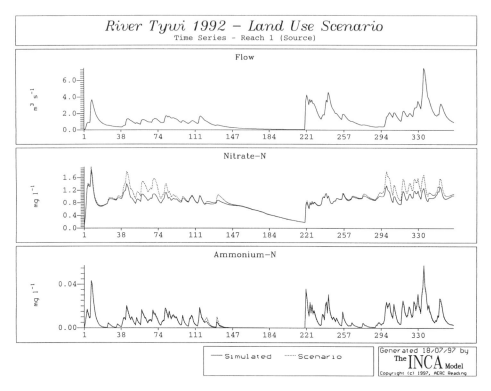

Figure 6.7   *Changing moorland land use in the upper reaches to fertilised grassland*

realistic land use change scenario the short vegetation (GNF) was altered to fertilised short vegetation. The effects of this are illustrated in Figure 6.7 which shows that fertilising the upland area would have a major effect on the nitrogen balance. Peak nitrogen levels are considerably higher as are mean levels during wet conditions.

By contrast, the effects of atmospheric deposition are minimal as shown in Figure 6.8. Increasing N deposition by 20% generates a small increase in nitrate-nitrogen in reach 1.

INCA has proved to be an extremely interesting tool for simulating catchment behaviour. It appears to reproduce the observed patterns of behaviour and changes in land use, hydrology and deposition, generating reasonable and acceptable results. The model is being used for more advanced scenario analysis and applied more widely in the UK and elsewhere.

ACKNOWLEDGEMENTS

This research has been supported by National Power and the Natural Environment Research Council. The opinions in the paper reflect solely those of the authors.

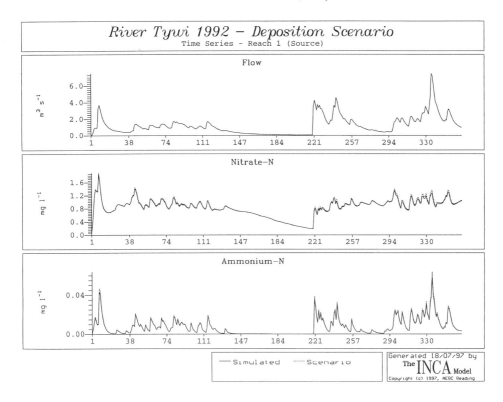

Figure 6.8   *Simulation of the Tywi with nitrogen deposition increased by 20%*

REFERENCES

Abbott, M.B., Bathurst, J.C., Cunge, J.A., O'Connell, P.E. and Rosmussen, J. 1986. An introduction to the Systeme Hydrologique Europeen (SHE). *Journal of Hydrology*, **87**, 45–77.

Addiscott, T.M. and Whitmore, A.P. 1987. Computer simulation of changes in soil mineral and crop nitrogen during autumn, winter and spring, *Journal of Agricultural Science, Cambs.*, **109**, 141–157.

Butterfield, D. and Whitehead, P.G. 1997. *The INCA User Manual.* Aquatic Environments Research Centre, Reading University.

Christophersen, N., Seip. H.M. and Wright, R.F. 1982. A model of streamwater chemistry at Birkenes, Norway. *Water Resources Research*, **18**(4), 977–996.

Cooper, D., Ragab, R. and Whitehead, P.G. 1993. *IHDM-TRANS A nitrate model for agricultural systems.* Institute of Hydrology report to the Ministry of Agriculture, Fisheries and Food.

Cosby, B.J., Wright, R.F., Hornberger, G.M. & Galloway, J.N. 1985a. Modelling the effects of aid deposition: assessment of a lumped parameter model of soil water and streamwater chemistry. *Water Resources Research*, **21**, 51–63.

Cosby, B.J., Wright, R.F., Hornberger, G.M. & Galloway, J.N. 1985b. Modelling the effects of acid deposition: estimation of long-term water quality responses in a small forested catchment. *Water Resources Research*, **21**, 1591–1601.

Cosby, B.J., Ferrier, R.C., Jenkins, A., Emmett, B., Tietema, A. and Wright, R.F. 1997. Model and Ecosystem and Loss of Inorganic Nitrogen (MERLIN). *Hydrology and Earth Systems Science*, **1**, 137–158.

DOE 1994. *Impacts of Nitrogen Deposition.* HMSO, London.

Edwards, R.W., Gee, A.S. and Stoner, J.H. 1990. *Acid Water in Wales.* Kluwer, London.

Green, F.H. and Harding, R.J. 1980. Altitudinal gradients of soil temperature in Europe. *Institute of British Geographers Transactions New Series,* **5**, 243–248.

Hall, R.L. and Harding, R.J. 1993. The water use of the Balquhidder Catchments: a process approach. *Journal of Hydrology,* **145**, 285–315.

Institute of Hydrology. 1996. *Hydromatic Register and Statistics.* Wallingford.

Jakeman, A.J., Littlewood, I.G. and Whitehead, P.G. 1990. Computation of the Instantaneous Unit Hydrograph and identifiable component flows with application to two upland catchments. *Journal of Hydrology,* **117**, 273–300.

Jenkins, A., Ferrier, R.C. and Cosby, B.J. 1997. A dynamic model for assessing the impacts of coupled sulphur and nitrogen scenarios on surface water acidification. *Journal of Hydrology,* **197**, 111–127.

Johnes, P.J. 1996. Evaluation and management of the impact of land use change on nitrogen and phosphorus load delivered to surface water: the export coefficient modelling approach. *Journal of Hydrology,* **183**, 323–349.

Lunn, R.J. 1995. *A Nitrogen Modelling System for Large River Basins.* PhD Thesis, University of Newcastle.

Meteorological Office 1981. *The MORECS System.* Hydrological Memorandum, No. 45.

Shamir, U. 1997. Hydrological basis for sound management of water resources. *British Hydrological Society Circulation.* **56**, 3–7.

Skeffington, R.A. and Wilson, E.J. 1988. Excess nitrogen deposition issues for consideration. *Environmental Pollution,* **54**, 159–184.

Whitehead, P.G. 1990. Modelling nitrate from Agriculture to Public Water Supplies. *Philosophical Transactions of the Royal Society of London, Series B,* **329**, 403–410.

Whitehead, P.G. and Williams, R. 1984. Modelling nitrate and algal behaviour in the River Thames. *Water Science and Technology,* **16**, 621–633.

Whitehead, P.G., Young, P.C. and Hornberger, G. 1979. A systems model of stream flow and water quality in the Bedford–Ouse River – 1. Stream flow modelling, *Water Research,* **6**, 1155–1169.

Whitehead, P.G., Beck, M.B. and O'Connell, E. 1981. A systems model of stream flow and water quality in the Bedford–Ouse River – 1. Water quality modelling. *Water Research,* **15**, 1157–1171.

Whitehead, P.G., Williams, R.J. and Hornberger, G.M. 1986. On the identification of pollutant or tracer sources dispersion theory. *Journal of Hydrology,* **84**, 273–286.

Whitehead, P.G., Williams, R. and Lewis, D. 1997. The QUASAR Water Quality Model. *Science of the Total Environment,* **194/195**, 399–418.

7

Basin-Scale Nitrate Simulation Using a Minimum Information Requirement Approach

Paul Quinn, Steven Anthony and Eunice Lord

INTRODUCTION

ADAS, as a part of UK consortium of land-based research groups, has participated in a project to investigate and ameliorate the losses of nitrate (N) from agricultural land. The principal goal is to determine which catchments and, in particular, which land use practices result in the EC limit of 11.3 mg/l NO_3-N within drinking water supply being exceeded. The project has been commissioned by the Ministry of Agriculture, Fisheries and Food and this paper reflects the ADAS component of this project.

The policy maker expects a scientist to arrive at a solution to the problems they pose. However, there are many ways in which the scientist can arrive at such a solution. The key question is which methods should be deemed as the best? From the outset the authors appreciate that there are many sources of uncertainty within the modelling system. In most cases it is not possible to remove this uncertainty, but it is the responsibility of the modeller to reduce the uncertainty to a minimum. The modeller should also critically report the sources and magnitude of any remaining error. Final model predictions must have a sufficient uncertainty estimate to allow a land use manager to design realistic policies. It is equally important to educate policy makers about the abilities and limitations of any modelling system and to produce tools that are useable and unambiguous. The ability to communicate realistic N estimates and their uncertainty to policy makers is a key obligation to scientists involved in catch-

Water Quality: Processes and Policy. Edited by Stephen T. Trudgill, Des E. Walling and Bruce W. Webb.

ment-scale modelling. It is also possible to target the sources of uncertainty so that informed decisions about science and policy issues can take place.

This paper describes models and data used in a prototype national N decision support system. Overall, the models and GIS data come together to give a methodology that represents the characteristic of dominant UK land use practices. A series of hydrologically and agriculturally similar classes have been established and predictions of flow and N can be made for catchments of any size.

Catchment-scale modelling requires the adaptation of complex, physical, "research" scale models in order to produce models that can run with the imprecise data we have available at the catchment scale. The research scale is treated as the experimental scale, that is, the laboratory, lysimeter, plot or hillslope scale. Complex physical models are regarded as representation of the observed processes seen in detailed experiments. However, the authors would argue complex descriptions of physical processes can *only* be used at this fine scale. Parameterising complex models at the catchment scale using large amounts of data introduces large uncertainty (Beven, 1989). The complex distributed nature of many models can make the uncertainty high due to the possibility of additive and mutliplicative error and all calibration is fraught with difficulties of equifinality (Franks *et al.*, 1997).

The MIR (Minimum Information Requirement; Anthony *et al.*, 1996; Quinn *et al.*, 1996) approach distils basic information and observed processes into a set of fundamental parameters and phenomena for the final model. In essence, MIR models use the minimum of amount of *available* data. A range of MIR models is needed within the final modelling system. It is part of the MIR approach that the model system can be broken down into a series of specific modelling components, all of which can be tested and emulated independently. In the current case we require three steps:

1. Processes that generate N in the soil available to leaching.
2. Processes that leach N from the soil.
3. Processes that move water and N through the catchment to our point of interest.

In this paper we describe how the more complex models are taken, studied in detail and then emulated with MIR models. The complex models have a full supporting literature which will not be discussed here; only the origin of the MIR will be described.

LEACHABLE SOIL NITRATE: EMULATING THE NITCAT MODEL

The NITCAT model (Lord, 1992) represents potential nitrate loss from a field as derived from available cropping information. The full model includes the field history (i.e. ploughed out grass, manure applications and previous crops), soil details, current crop, and current husbandry. The model is essentially a N balance model and takes into account inputs and outputs as modified by nitrogen increase in autumn, mineralisation of nitrate and residues from previous crops. The model is essentially a data-driven model based on extensive field work. The experiments show the importance of N loss from pig and poultry manure, intensive dairy farming and from potatoes and oil seed rape. The model has been tested and evaluated within 10 Nitrate Sensitive Areas (NSAs; Lord, 1992), where all the model inputs were available and *in situ* validation measurements were taken.

Two components have been retained from the NITCAT model for our MIR model operation. The first is that the N available for leaching increases slowly until a maximum crop yield is reached, after which large amounts of excess N will occur (Figure 7.1). Secondly, the excess N available for leaching can be determined for a range of crops (Table 7.1).

An estimate of the likely value for the N available for leaching can be determined using Monte Carlo simulations and from the abundant data relating to farmer practice. The result of this analysis is a clear description of the range of possible N values available to be leached in one year. Representation of the variability in N availability is also important. Individual fields may have different leachable N values but if a statistical distribution of these values is made for a larger area (a catchment or region) then a good estimate of the distribution of N available to leaching can be made. Thus, we have an estimate of our input uncertainty that we can propagate through to the final model predictions.

Figure 7.2A, shows a typical distribution of fertiliser N inputs for winter wheat for

Table 7.1 *Some example crops and their mean soil N content as a function of preceding crop N*

Crop	N available for leaching (kg/ha)
Winter wheat	35
Winter barley	35
Potatoes	80

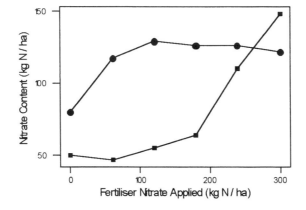

Figure 7.1 *The N balance for a typical crop showing the nature of N available for leaching. Circles are the crop N content and squares are the soil N available for leaching*

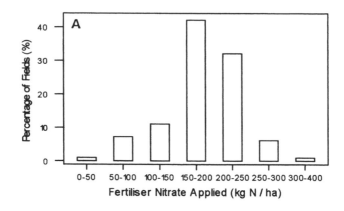

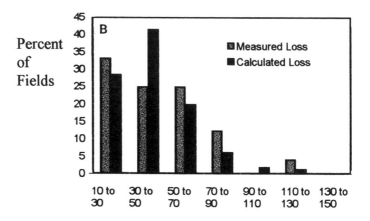

Leached Nitrogen (kg / ha)

Figure 7.2   *(A) The distribution of fertiliser N applications to winter wheat within the UK. (B) Results from NITCAT compared to measured values for winter wheat*

the UK situation. Figure 7.2B, shows the amount of N likely to be leached from winter wheat in one growth year. The results from the NITCAT model are also compared to the measured N loss derived from winter wheat fields within the NSA scheme. It can be seen that the distribution of likely N losses is different from the distribution of the fertiliser inputs. The total N available to be leached per crop is all that is required to simulate N content within river catchments (as will be demonstrated later). The NITCAT model and its fundamental data sets have been used to their full potential off-line, but only the minimum amount of information is retained for later use.

NITRATE LEACHING: EMULATING THE SLIM MODEL

The SLIM model (Addiscott *et al.*, 1986) allows a full physical description of soil, husbandry and meteorological inputs. The model is one-dimensional and allows distributions of N across the soil profile. N loss can be calculated for any layer. Hydrological processes are controlled by meteorological inputs and by crop growth dynamics. N loss is related to the total N in the profile and the soil type. In the analysis of the model output, it was assumed that water and N exit the soil at a depth of 90 cm (i.e. at a depth similar to field drains and coincidental with the depth of suction lysimeter measurements). Numerous simulations of the model, for different combinations of crops, initial N distribution and weather conditions, were accumulated. Distinct patterns were seen for the cumulative N loss as a proportion of the total drainage water. It is possible to plot the cumulative N loss as a proportion of the total N available for the leaching process (as described above). Thus if N loss was plotted against a drainage efficiency term a clear pattern was seen. The drainage efficiency term (ε) is defined as the cumulative drainage divided by the soil water-holding capacity of the current soil (to a depth of 90 cm). The results showed that the proportion of N leached is the same rate for all soil textures, for any value of ε and any initial N content. Figure 7.3 shows the result of SLIM model for three soil types showing that the normalised outputs lie along a single line.

It is now possible to use a simple MIR model to represent the full output dynamics of the model. The solid line on Figure 7.3 is a best-fit curve to all the data. The curve has the function:

$$\varepsilon = \text{HER}/\theta$$
$$P = 1.1111\,\varepsilon - 0.203\,\varepsilon^3 \qquad \text{where } \varepsilon < 1.33$$
$$P = 1.0 \qquad\qquad\qquad\quad \text{where } \varepsilon > 1.33$$

Figure 7.3 *The proportion of total N leached as a function of drainage efficiency for sand (squares), silty loam (triangles) and silty clay (circles). The symbols are the output of the complex model and the solid line is an MIR mimic of these results*

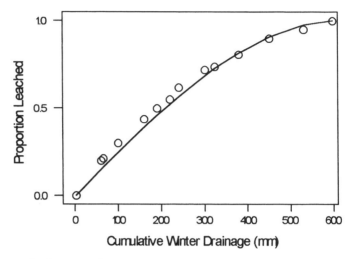

Figure 7.4   *Graph of measured (circles) and simulated N leached for a clay loam on grazed grass. After Scholefield et al. (1993)*

HER (Hydrologically Effective Rainfall) is the cumulative soil drainage, θ is the soil water-holding capacity (for 90 cm) and P is the proportion of available N lost.

This MIR requires an external input of HER (which is shown later). The soil water-holding capacity term is a commonly used soil characteristic and can be estimated from a knowledge of the soil class. The MIR model can be tested against field data in the same way as the SLIM model as can be seen in Figure 7.4.

HER GENERATION: EMULATING THE IRRIGUIDE MODEL

IRRIGUIDE (Bailey and Spackman, 1996) is a commonly used crop water use model that simulates soil moisture deficit and HER for any location in the UK (using any one of 60 UK synoptic daily weather stations). The most important component of the model is that it can grow crops in response to current meteorological and soil conditions. The model can run for any popular UK crop or soil and has been extensively tested. The model is similar to the MORECS model (MORECS, 1996) approach but has attempted to extend the range of crops and improve the dynamics of crop growth.

The model is one-dimensional and is essentially a soil moisture deficit model. Drainage of water through the profile occurs after soil field capacity has been reached. A daily time step is used; Soil Moisture Deficit (SMD) and HER are calculated from a simple water balance equation.

HER is generated as excess vertical flow when an SMD of zero is exceeded. A correction for temporary storage and infiltration on the surface of soils on flat land is allowed. Firstly, an estimate of Reference Potential Evapotranspiration (RPE) from the Penman–Monteith equation (Monteith, 1965) is made. RPE is assumed to be an unstressed wet grass. A modification to the RPE is made for each specific crop to allow for the age and size of that crop. Simulations have shown that deviations from the RPE occur when crops are first sown and as the crops mature. The current MIR approximation uses the crop potential evaporation (CPE) to drive a simple one-parameter bucket storage to simulate SMD and HER. The MIR requires an assumption of a typical "average" husbandry practice for each specific class of crop. For example, the winter wheat class used in this study assumes the crop is planted on 1 October and is harvested on 1 September (with one month of fallow ground). Thus all framers are assumed to be behaving in a similar way. A perennial grass would be grown continuously across the year with 100% crop cover. An urban class has been allowed that has a shallow effective routing depth (mimicked here by running a fallow cover with a shallow routing depth).

The system therefore needs a time series of CPEs for every land class allowed in the model system. The CPE generator can be built into the final model system and can be activated when a catchment is selected.

The size of the storage bucket used for SMD and HER is taken form a look-up table of parameters for UK crops. The maximum rooting depth is the only value needed. This value is combined with the water-holding characteristics for the current soil type. IRRIGUIDE allows two water availability values: (1) "easy" water, and (2) "restricted" water. "Easy" water is extracted for the most part and only under extreme drying conditions is extra "restricted" water released. The value taken for our MIR is the average of the two values. Therefore a Medium Sandy Loam from IRRIGUIDE has values of: easy = 16% and restricted = 18%. So a crop with a maximum rooting depth of 1 m would allow 160 mm of easy water and 180 mm of restricted water. The value used for the capacity of our MIR bucket is taken as 170 mm of water. Water is extracted from the bucket at the full CPE rate until empty, when evaporation stops.

A hydrological model called TOPCAT-N (Quinn *et al.* 1996, see below) contains the bucket MIR. TOPCAT-N receives the inputs of the CPE and rainfall and calculates the HER time series per crop. The land classes used in this paper were tested against the complex model (IRRIGUIDE) and against observed field data (Figure 7.5). Figure 7.5A shows that differing crops generate differing HER amounts and that the shallow rooted class can generate substantial amounts of HER. Variations in total HER are effected by two components: (1) the remaining SMD caused by the previous summer's drying, and (2) the difference in CPE between newly planted and perennial crops.

Figure 7.5B and C shows that the MIR has a near-perfect fit to the IRRIGUIDE simulations. This is due to the fact that both IRRIGUIDE and the MIR are controlled by the same CPE and rainfall, and by the total water available to evaporation which in most years is not exceeded. Figure 7.5D shows a test against an independent time series of HER taken from the Woburn Sands experiment (where the MIR was set up for the same conditions as the experiment). The conclusion of this work is that a good quality CPE generator or CPE MIR is needed by the system. The use of crop-specific potential evaporation must be retained as there are a number of circumstances that can seriously influence HER generation: (1) during the establishment of winter crops, and (2) during the establishment of spring crops. Commonly the total amount of precipitation and its

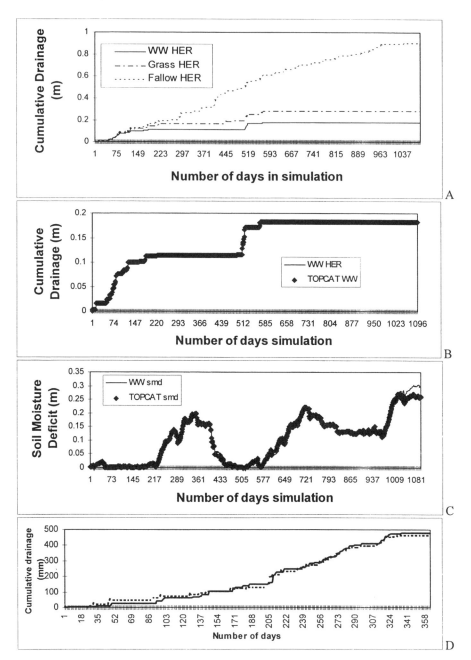

Figure 7.5 *(A) HER time series for a three year daily simulation for Grass, Winter Wheat (WW) and Fallow (Fallow is assumed to be equivalent to the Urban class). (B) HER predicted by TOPCAT and compared to the HER from IRRIGUIDE. (C) SMD predicted by TOPCAT compared to the SMD from IRRIGUIDE. (D) Observed and simulated HER from TOPCAT compared to the Woburn Sands experiment*

frequency do not create great crop stress during these periods; however, when crop stress occurs it can suppress crop development and alter the CPE rates.

CATCHMENT HYDROLOGY: EMULATING TOPMODEL

TOPMODEL (Beven and Kirkby, 1979) is a quasi-physically based runoff generation model that, in its full mode of operation, uses a topographic index to control the depth and activity of the water table. In our case a simple MIR catchment version of TOPMODEL has been developed called TOPCAT. The aim of this MIR is to route HER to the outfall and simulate the attenuation of flow for any size of catchment. The model is thus representing both the hillslope and the channel component of the flow routing. In essence the model describes the response time of a catchment and its recession characteristics. TOPCAT utilises the same subsurface flow equation as TOPMODEL and capitalises on the model's ability to represent the total wetness of a catchment and its relationship to flow. The bucket MIR described above feeds the HER time series to the routing equations. Flow (and its associated N concentration) is then routed to the outfall using a scaled recession parameter (m). When HER is generated from any individual land class it recharges the total catchment wetness term and then the subsurface flow is removed from the catchment:

$$SBAR_t = SBAR_{t-1} - QIN + QB$$

and

$$QB = Q_0 \, e^{(-SBAR/m)}$$

where $SBAR$ = the current catchment soil moisture deficit, QIN = the calculated total of HER per land class multiplied by the fraction of that land class, QB = the subsurface flow exiting from the catchment soil moisture store, Q_0 = upper rate of flow for a catchment under very wet conditions (this can be set to an average value taken from TOPMODEL simulations, and m = the recession rate parameter.

The "activity" term m in our prototype model could be linked to the HOST (Hydrology of Soils Types) classification (Institute of Hydrology, 1995) and the area of the catchment. This is the subject of recent research. Thus, current simulations have required at least a limited period of observed flow data. However, for the UK, the ability to assign a reasonable m parameter to a catchment is not a fundamentally difficult task. If the user can acquire some flow data from the stream network then m can be calculated using a relationship of $1/Q$ against time (Beven *et al.*, 1996).

If a time series of rainfall and a suite of CPEs are input, a single time series of HER can be produced for the whole catchment. TOPCAT then mixes all the flow together assuming that the flow has been generated per unit length of channel.

AN APPLICATION TO THE RIVER GREAT OUSE

The River Great Ouse is a surface water catchment supplying a water intake at the Clapham intake (1400 km^2). The area is intensely agricultural and contains some large urban areas. The Great Ouse has been designated a Nitrate Vulnerable Zone (NVZ) and will thus be subject to future land use regulations in order to reduce N inputs from agriculture.

Figures 7.6 to 7.9, show a range of data that are used to drive the model taken from the ADAS national GIS. When the user selects a catchment, all the data required to run the TOPCAT-N are returned automatically. Given our knowledge of the uncertainty in the data sets we can also set realistic worst-case scenarios for the model. Figure 7.6 shows The River Great Ouse catchment and subcatchments, the river network (the subcatchment numbers are used in later analysis). The catchment boundaries were defined by the old National Rivers Authority (as such all the subcatchments have a range of useful flow statistics).

Figure 7.7, shows the recorded soil type per 1 km. In our prototype system only one effective soil type is allowed for the whole catchment therefore the GIS data needs to be pre-processed. Table 7.2, shows the distribution of soil classes per subcatchment and also estimates the average value per subcatchment and the final value used in the MIR model.

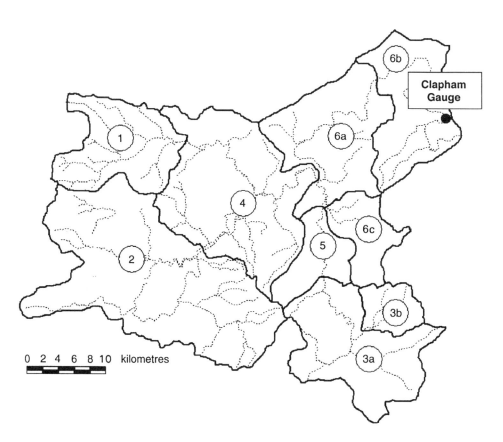

Figure 7.6   *The River Great Ouse, internal subcatchments and the location of the Clapham Gauge*

Table 7.2 *Soil texture distributions for the subcatchments shown in Figure 7.7. The final value used in the MIR model is shown in bold*

Subcatchment	Clay (km²)	Clay Loam (km²)	Silty Loam (km²)	Sandy Loam (km²)	Sand (km²)	Total (km²)	Weighted average
1	64	0	17	17	0	98	43.51
2	106	30	10	186	0	332	37.85
3a+3b	93	2	7	30	7	139	42.50
1+2+4	338	30	40	243	0	651	41.00
3a+3b+5	108	2	7	53	9	179	41.30
ALL (1+2+3+4+5)	630	32	52	328	12	1054	**41.96**
Water-holding capacity (%)	48.3	40.5	37.5	31.5	18.3		

Soil class

- ■ Clay
- ▦ Clay loam
- □ Loamy sand
- ▨ Silty loam
- ▩ Loam
- ▥ Sandy loam

Figure 7.7 *The spatial distribution of dominant top soil textures. Data reproduced under licence from the Soil Survey Land Research Centre within the UK national nitrate modelling project NT1701 funded by the Ministry of Agriculture, Fisheries and Food*

The value of effective average water-holding capacity estimated for the whole catchment is equivalent to Clay Loam, although the catchment is dominantly Clay. However, we know from expert knowledge of the area that cultivated Clay soils will contain land drains and mole drains and the effective field capacity is probably lower than that suggested by the GIS. So, a lower water-holding capacity is probably more realistic. Therefore, it is quite important to deduce a reasonable value for effective soil type based on both soil class and on husbandry expertise. We should also appreciate that this incurs uncertainty and should be added to our final predictions. The GIS can tell us much, but it should only be used as a guide to the deduction of a final distribution of parameter inputs.

Figures 7.8 and 7.9 show just two layers from the national GIS database of agricultural statistics. The first class reflects all the arable land (which is mainly winter wheat and winter barley) and the second is the agricultural grass (which also contains the effects of animals on the total N value). Another important assumption of the model is that the subcatchments are agriculturally similar. Hence, flow and its associated N content would be produced equally throughout the catchment. If this

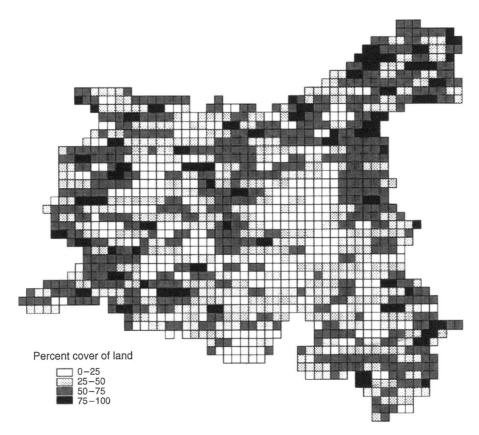

Figure 7.8   *The spatial distribution of arable land as a percentage of all land (within a 1 km cell)*

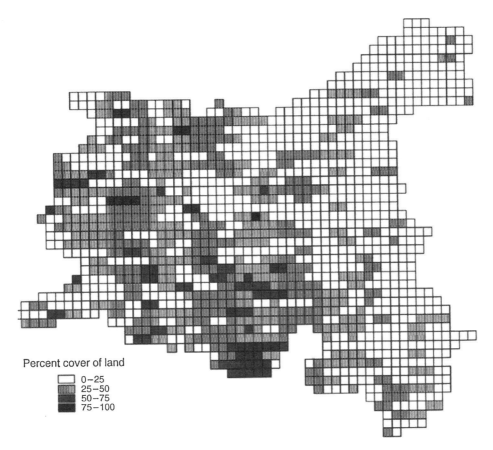

Figure 7.9 *The spatial distribution of grassland as a percentage of all land (within a 1 km cell)*

Percent cover of land

- 0–25
- 25–50
- 50–75
- 75–100

assumption holds true then a simple statistical representation of land use patterns can be used and there is no need for a distributed model (this allows a huge reduction in operational model complexity). To demonstrate this assumption the catchment statistics were derived for Figures 7.8 and 7.9 as well as urban areas. Table 7.3 shows the results of the land class distributions.

In the simulations shown below we have allowed three land classes.

1. The arable class, which has been approximated to by a winter wheat (sown on 1 October and harvested on 1 September).
2. A perennial grass class, which is assumed to have full crop cover all year round. A component of the N total includes the effects of animals kept on that land.
3. An urban class with low effective rooting depth (N availability taken as 0 kg/ha as a value of urban N is represented by a constant N concentration set for the background dry weather flow rate).

Table 7.3 *Distribution of land use for the Ouse subcatchments. The total remaining area is made up of woodland and water-covered areas*

Subcatchment area	Arable (%)	Grass (%)	Urban (%)	Total area remaining (%)
1	44.07	35.22	1.59	19.12
2	43.36	42.81	2.32	11.51
3a+3b	49.53	35.36	7.47	7.64
1+2+4	44.60	38.00	4.17	13.23
3a+3b+5	45.62	34.31	10.56	9.51
ALL 1+2+3+4+5+6	48.07	34.59	5.69	11.65

Whilst more classes will be required within the final system, it was of interest to find out how much accuracy and uncertainty can be represented with these three classes alone. In our simulations the absolute areas involved per land class are taken as correct. The meteorological inputs are also generated as a single time series for a position representing the centroid of the catchment. A set of hydrological catchment parameters has been set using a short period of flow values taken at the outfall of the catchment. In essence this is a hydrological flow calibration exercise. A validation of the hydrological model was made using measurements at the Clapham Gauge. Figure 7.10 shows the observed and simulated flows. A background flow rate is required and a constant N concentration of 4 mg/l has been allocated to this flow.

It would be possible to build in our uncertainty in our land class estimates, error in the meteorological data allow for variation in planting and harvest dates. For the moment, a full uncertainty analysis has not been performed. However, the findings of our first uncertainty studies are still quite revealing.

LIVING WITH UNCERTAINTY

In all our analyses we are keen to show which parameter inputs are the most sensitive to our final system. The parameters remaining in our model system are:

1. The total available N to the leaching process at the onset of winter drainage.
2. The effective water-holding capacity of the soil.
3. The area per land class for the whole catchment. Also the number of land classes needed.
4. The accuracy of the input meteorological data.
5. The crop husbandry actions and the depth of crop roots.

It became quite clear that the greatest source of uncertainty comes from our estimate of the N available for leaching term. The effective soil water-holding capacity is less sensitive than the N term but in relative terms is more sensitive than the others. The area per land class is a reasonable representation of the land cover within the catchment; however, when creating new land use scenarios a good estimate of uncertainty may be needed. The meteorological data may have error but the range of events is a statistically realistic climate scenario for the catchment. Whilst one of our

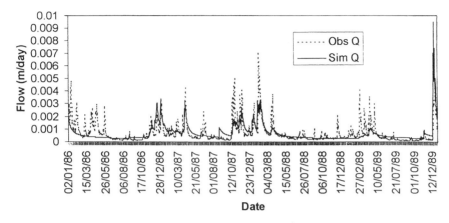

Figure 7.10   *The observed and simulated daily discharge at the Clapham Gauge following a hydrological calibration exercise*

modelling goals is to check whether reasonable fits to N data can be achieved, it is equally important to show the variability of our predictions in the light of parameter uncertainty. The crop husbandry and rooting depths have been taken as average UK values.

The following simulations will show the mean, best and worst-case simulation results (as seen in Figure 7.11). The mean result is a simulation returned automatically from the modelling system where the N available for arable land is 45 kg/ha and for grass cover 55 kg/ha and there in an effective soil texture of Clay Loam. The worst-case scenario is arable N leaching at 55 kg/ha and grass at 65 kg/ha and the effective soil water-holding capacity was set for Sandy Loam (300 mm per 1 m of soil). The best-case scenario uses 35 and 45 kg/ha for arable and grass with a soil texture of Clay (400 mm per 1 m of soil).

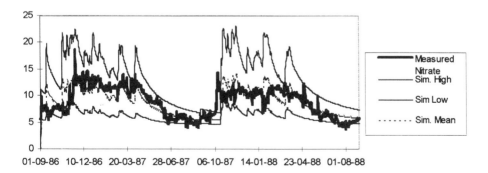

Figure 7.11   *A two year daily simulation from the model system (1986,87), containing the automatic output from the model (the simulated mean) as well as our two extreme simulations (simulated high and low)*

The results clearly show that with the leachable N and soil texture terms alone we can seriously affect output from the model. This implies that the estimate of the uncertainty in these terms is paramount to the final model system. It can be argued that these two terms outweigh, or even drown out, the uncertainty of the other terms. Future work will analyse the relative magnitude of our uncertainty for all model components. The research problem remains, that improved catchment-scale N and soil texture estimates are needed within our models. An ultimate solution to this problem may take a long time to achieve. Thus, an immediate research need is to estimate the risk of occurrence of error in our model predictions. The large error bars on Figure 7.11 may seem bleak but it must be noted that the chance of such extreme circumstances occurring will be rare. So it can be concluded from this work that it is the realistic likelihood of occurrence of a simulation that is the key to progress in large-scale water quality modelling and policy making. Also on Figure 7.11 we can see that the mean simulation is quite close to the observed N concentrations; this reflects that the basic information used in the simulation may be quite representative of the catchment. Simulations are more likely to give results closer to the mean result, but the likelihood of this occurrence needs to be explicit.

A final point on the N simulation is that a good fit to the observed N concentrations can be achieved if calibration of the N leaching parameters is allowed.

CONCLUSIONS

This paper has attempted to discuss a number of issues related to catchment-scale water quality modelling for policy support.

- A series of simple, elegant catchment water quality modelling tools, fed by available GIS data, has been presented. The MIR approach is proposed as a means of making the final operational model simple, useable and capable of representing our uncertainty.
- The MIR can emulate a complex model and be validated in the same way.
- A full evaluation of the variability in our key model parameter inputs is made and their effect on model output has been shown.
- Uncertainty analysis passes across to the user a realistic range of results that allows a more informed set of policy decisions to be determined. Moreover, by ranking the sources of uncertainty we can see where our future work lies in reducing that uncertainty.
- A tool for estimating catchment N losses, that can also reflect land use changes has been described.

It may seem a deterministic "ideal" to use fully physically based models with all the GIS data available for large catchments. However, can such complex models be justified when our ability to even measure the mean effective value of leachable N remains so weak. MIRs rely on field experiments and physically based models for there construction; they do not supersede these models, rather they show their appropriate role in the catchment modelling scheme. The final decision support system will include a number of physically based models that can be run off-line. However, users will be forced to switch model structure when they wish to simulate effects at the catchment scale. Physically based models and experiments are our archive of knowledge on processes. The MIRs are used to construct the operational, useable model system. Despite the complexity of processes occurring, simple MIR models can reflect the dynamics of catchment-scale measurements. GIS is a powerful tool for analysis giving a vast amount of information at our fingertips. All the information together can give a good guide to N loss with time and for future land use change; however, we will not get an exact estimate of N loss.

REFERENCES

Addiscott, T.M., Heys, P.J. and Whitmore, A.P. 1986. Application of simple leaching models in heterogeneous soils. *Geoderma*, **38**, 185–194.

Anthony, S.J., Quinn, P.F. and Lord E.I. 1996. Catchment scale modelling of nitrate leaching. In Quinn, P.F. and Lord, E.I. (Eds), *Modelling in Applied Biology: Spatial Aspects*. Aspects of Applied Biology No. 46, 23–32. Association of Applied biologists, Wellesbourne, UK.

Bailey, R.J. and Spackman, E. 1996. A model for estimating soil moisture changes as an aid to irrigation scheduling. 1) Historical background, operational details, model description and sensitivity. *Soil Use and Management*, **12**, 122–128.

Beven, K.J. 1989. Changing ideas in hydrology – the case of physically based models. *Journal of Hydrology*, **105**, 157–172.

Beven, K.J. and Kirkby, M.J. 1979. A physically-based variable contributing area model of basin hydrology. *Hydrological Science Bulletin*, **24**, 43–69.

Beven, K.J., Lamb, R., Quinn, P.F., Romanowicz, R. and Freer, J. 1996. TOPMODEL. In Singh, V.P. (Ed.), *Computer Models of Watershed Hydrology*. Water Resources. Publications, Littleton, Colorado, USA.

Franks, S.W., Beven, K.J., Quinn, P.F. and Wright, I.R. 1997. On the sensitivity of soil–vegetation–atmosphere transfer (SVAT) schemes: Equifinality and the problem of robust calibration. *Agriculture For Meteorology*, **86**, 63–75.

Institute of Hydrology. 1995. *Hydrology of Soil Types*. Wallingford.

Lord, E.I. 1992. Modelling of nitrate leaching: nitrate sensitive areas. In Nitrate and Farming Systems. Aspects of Applied Biology No. 30, 19–28. Association of Applied Biologists, Wellesbourne, UK.

Monteith, J.L. 1965. Evaporation and environment. *Symp. Soc. Exper. Biol.*, **19**, 205–234.

MORECS 1996. *Meteorological Office Rainfall and Evaporation Calculations Scheme, Version 2*. Meteorological Office.

Quinn, P., Anthony, S., Lord, E. and Turner, S. 1996. Nitrate modelling for the UK: a Minimum Information Requirement (MIR) Model Approach. *Inter Celtic Symposium*, 8–11 July 1996. INRA Publications No. 46.

Scholefield, D., Tyson, K.C., Garwood, E.A., Armstrong, A.C., Hawkins, J. and Stone, A.C. 1993. Nitrate leaching from grazed lysimeters: effects of fertilizer input, yield drainage, age of sward and patterns of weather. *Journal of Soil Sci.*, **44**, 601–613.

Section Three

LINKING SCIENCE AND POLICY

8

Bridging the Gap Between Science and Management in Upland Catchments

Pauline E. Kneale and Adrian T. McDonald

INTRODUCTION

Nineteenth century catchment management in the north of England evolved to supply high quality water to expanding cities. Cities like Leeds and Bradford had their own Water Boards, each developing supplies from Pennine sources, building reservoirs, aqueducts, treatment works and distribution infrastructures. To ensure drinking water quality they looked to four main lines of defence: pristine catchments, long storage times in large reservoirs, filtration and disinfection.

Over the past century these defences have been eroded. Pristine catchments succumbed to pressures on the land. The essentially unproductive and unpopulated catchments of the early nineteenth century have been subject to agricultural improvement, game management and vastly increased recreational access. Forestry provided increased income and as the Water Boards were amalgamated, and then privatised, the land was sold or passed to tenant farmers. Reservoir storage times are typically lower today than in the last century owing to increased demand. There is greater awareness of the presence of dead zones and the effect of bypassing which leads to very variable water residence times within reservoirs. Where bypassing occurs there may not be time for micro-organisms to decay while in storage. Filtration and disinfection are still critical defences for potable water. With increasingly diverse uses of catchments, the degree of chemical treatment has increased and new problems arise. New

Water Quality: Processes and Policy. Edited by Stephen T. Trudgill, Des E. Walling and Bruce W. Webb.
© 1999 John Wiley & Sons Ltd.

understanding of chemical interactions and links to medical effects has changed appreciation of treatment processes. Chlorination to disinfect coloured waters for example may produce trihalomethanes (Casey and Chua, 1997; Garcia-Villanova *et al.*, 1997; Williams *et al.*, 1997), compounds which have been implicated in carcinomas (Koivusalo *et al.*, 1997) and possibly with reproductive and development effects (Reif *et al.*, 1996). The number and percentage of water supply samples contravening the permitted concentration values (PCV) in the Yorkshire Water region for 1995 is given in Table 8.1, showing, for example, that 3.1% of samples have trihalomethanes above the guideline.

Scientific understanding of the relationships between water characteristics, climate and source areas is advanced enough to enable evaluation of different management strategies. These evaluations require additional understanding of risk, law, public interest and economics. In practical catchment management all of these issues impinge on decision making and it is this multifaceted management problem that we explore in the context of potable supply from rivers and reservoirs. In Yorkshire, as in much of the United Kingdom, rivers are fundamental to potable supply and we focus here on river catchments, but many of the issues also relate to groundwater catchments.

Recently the expected role of management, the provision of water to meet all demands, has changed as the water resource has become less assured. Demand from industry may be effectively reduced through pricing policies, agricultural abstraction licences are granted subject to supply above a specified minimum threshold in the river or borehole to safeguard downstream abstractions and meet "in river" needs (Environment Agency, 1997). The drought of 1995 and increasing pressure to maintain flows in rivers such as the Wharfe at ecologically desirable rates, has raised public awareness of

Table 8.1   *Water quality in supply zones. Data from Yorkshire Water Services Limited (1995)*

Parameter	Total	Contravening PCV	
		No.	%
Coliforms	12 483	90	0.7
Faecal coliforms	12 482	11	0.1
Colour	12 474	1	< 0.1
Turbidity	12 475	45	0.4
Odour	1989	0	0.0
Taste	2004	0	0.0
Hydrogen ion	12 474	19	0.2
Nitrate	2100	14	0.7
Nitrite	2239	5	0.2
Aluminium	6601	18	0.3
Iron	6619	184	2.8
Manganese	6600	71	1.1
Lead	2411	85	3.5
PAH	1110	34	3.1
Trihalomethanes	1564	48	3.1
Individual pesticides	107 708	118	0.1
Other parameters	30 544	8	< 0.1

water supply issues. Ofwat is keen to promote price-based demand management but is at the same time promoting increased resource-demand margins. Ofwat are recognising a factor to account for the resource-demand margin for supply security as part of the next price review assessment.

MANAGEMENT ISSUES

Meeting Increased Demand

Development of upland impoundments is one of the most effective ways of providing a clean, efficient, reliable and cheap water supply. Public reaction to the development of a new supply source is seldom supportive, owing in part to the fact that it may take 10 or more years to agree, purchase, plan and develop a new reservoir and at the time when the new supply need is first mooted, the need is simply a forecast and as such can be criticised and condemned as unnecessary. Recognising this, water companies and their predecessors in England and Wales have expanded the gathering grounds while leaving engineering works essentially unchanged. In this way they can cope with increased demand without the environmental impact outcry that inevitably follows a major engineering development in the uplands. This strategy is not without an impact on storage characteristics in these resource systems which makes the overall system more prone to failure during drought.

On some catchments a choice exists about whether to accept or reject waters into the reservoir. On direct gathering ground the existence of a bypass channel, often created as part of the engineering works required as part of the impoundment creation, determines whether water must be taken into the reservoir. The presence and the nature of catchwater systems determine the flexibility or otherwise of choosing waters that come from the indirect catchment area. In the example developed below and outlined in Figure 8.1 the Inmoor and How Stean catchments, some 26 subcatchments in total, are individually directed by variable-height weirs. The Inmoor system has plank gates which are manipulated by hand. The How Stean subcatchments are controlled by screwgate weirs operated by one person and these could be automated, although power provision and signal transmission problems and costs are currently non-trivial issues. The scientific manipulation of such catchwater systems, to deliver or exclude particular waters, depends upon an understanding of the raw water qualities that are likely to arise from specified parts of the catchment under particular circumstances.

There are three types of upland gathering ground:

- direct impounding
- indirect impounding
- indirect non-impounding.

Each offers the manager a different portfolio of possible management options (Table 8.2). The gathering grounds in the upper Nidd valley, developed originally for Bradford Corporation Waterworks, are a network of reservoirs, intakes to inter-catchment

Table 8.2 *Management strategy options in catchments*

	Direct impounding	Indirect impounding	Indirect non-impounding
Land management	✓	✓	✓
Catchment manipulation and selection		✓	✓
Storage management	✓	✓	

transfer pipes, and river and aqueduct transfers. The upper part of the system is shown schematically in Figure 8.1.

In the upper catchment, flow to the reservoirs is natural, and the supply is augmented by gravity-fed flows from the Inmoor and How Stean catchments in the adjacent valley. The catchments are direct, and the flow is normally sequential. Waters derived from these areas flow naturally by gravity to the impoundments. The bulk of the indirect catchments, that is catchments in which the water would naturally bypass the impoundment, were added as a third phase at a later date. Finally, along the route of the 50 km aqueduct to the treatment works, a number of sources, some themselves the products of direct impoundments, discharge waters directly to the aqueduct. Such flows are more difficult to manipulate and may effectively displace aqueduct waters already secured and managed upstream. These differences arise largely from the political expediency of being able to develop a source without generating controversy, but they offer to the catchment manager different management opportunities. To control raw water quality the manager has three broad options, namely:

- land management
- catchwater manipulation
- storage controls.

Not all of these will be available in each of the three catchment types (Table 8.2).

Security

Managers need a measure of the security of a catchment. This may be done by identifying where the catchment lies on the continuum from fragile to robust. This can be achieved by examining five supply system attributes (Figure 8.2).

1. Land use and access vulnerability requires assessment of individual processes and cumulative evaluation to calculate overall risk. For example tanker spill risk depends on length, grade and usage of public highways in the catchment; colour depends on the presence of particular organic soils and connectivity with watercourses; *Cryptosporidium* depends on stocking density at key times of the year.
2. Land management potential based largely on ownership and management status:

   - owned and managed

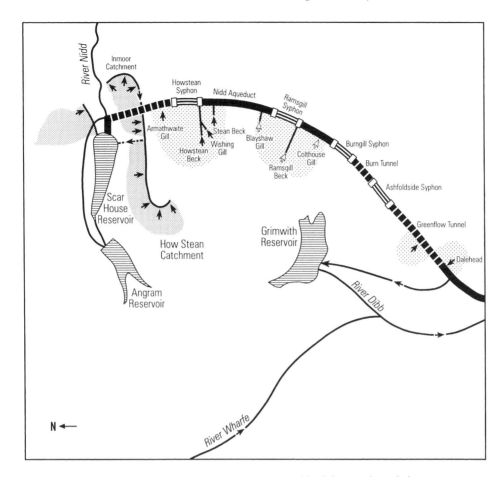

Figure 8.1 *Integrated distribution system in Upper Nidderdale, North Yorkshire*

- owned but tenanted
- not owned but effective management agreement in place
- not owned, no management guidelines.

3. Ability to manipulate catchwaters or storage, dependent on catchment type outlined above.
4. Water treatment plant capability which is dependent on capacity, process duplication, "polishing" and the availability of alternative feedstocks.
5. Demand site treated water source substitution and grid position.

A catchment at one extreme might be classed as "robust", having low problem potential, owned and managed by the company through a catchwater system open to manipulation, leading to a high quality treatment works supplying a town for which alternative water feed stocks are available. At the other extreme a fragile catchment might have a problematic colour or *Cryptosporidium* problem, be neither owned nor

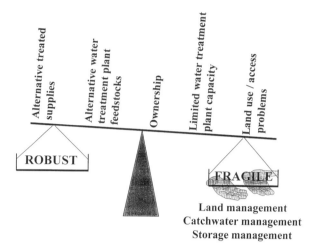

Figure 8.2   *Factors influencing the robustness and fragility of catchments*

managed by the company, have waters that cannot be diverted, have a treatment facility of limited capacity and supply a town which has no alternative means of water supply. Such a bleak management scenario is not unreal. Northallerton, North Yorkshire, is supplied from Osmotherly reservoir which has a road along its northern edge. Potentially this supply is at risk from spills or damage following public access. The catchment is only partially owned by the water company and there is evidence of increasing vandalism, particularly of bonfires which increase the risk of fire to forest on the south side of the reservoir. The treatment works lack duplication and the town has no more than a four inch supply support from the grid. Currently work is underway to improve links to the grid, improve the fire problems and control forest harvesting which could generate excess sediment and coloured runoff.

Catchment fragility can be identified where loss of surface vegetation cover and consequent erosion bring problems with increased sediment, Al, Mn, Fe and colour in water, and the impoundment capacity of reservoirs is decreased (White *et al.*, 1996). Vegetation cover on peat catchments, once broken for example by fire, trampling, over-grazing, ditching or during tree harvesting, is very hard to re-establish. Downcutting by water quickly establishes a gully that is often deemed to be self-perpetuating. Recent work by the Tweed Foundation has demonstrated, however, that the exclusion of grazers rapidly promotes vegetation cover along the channel (Dr R. Campbell, Tweed Foundation, pers. comm.).

Storage times

The mathematical concept that reservoir storage, *S*, divided by the daily outflow volume, *Q*, gave a number of days' storage depends on an orderly queuing characteristic most water systems simply do not possess. Yorkshire Water amongst others has shown considerable short-circuiting of inflow waters across reservoirs. At Scar House and Angram Reservoirs (Figure 8.1), harmless but uniquely identifiable bacterial phage tracers were placed on the catchments immediately beside streams that fed into catchwaters. The tracers passed along the stream and catchwater, fed through a tunnel to the reservoir, across and down approximately 35 m to the aqueduct intake in the base of the dam and thereafter travelled 50 km to the Chellow Heights treatment works where they were detected. In total the travel time was consistently below 36 hours when the theoretical storage is approximately 150 days (McDonald, 1988).

Cryptosporidium

The source of *Cryptosporidium* outbreaks affecting potable supplies can be a point such as a slurry store, or non-point from grazing land or after muck spreading. *Cryptosporidium* is difficult to detect and very resistant to treatment. Unlike bacteria which are unlikely to survive upwards of a few weeks on a catchment, *Cryptosporidium* oocysts can remain viable for three years (Hunter and McDonald, 1991). The long-lived nature of the oocyst means that once washed off the catchment into a river it may remain dormant in sediments for considerable time periods and be remobilised during later storms (Marshall *et al.*, 1995).

An effective treatment for *Cryptosporidium parvum* in potable supplies is a research focus world-wide. Oocysts are generally resistant to disinfectants (Fayer and Ungar, 1986) and chlorination at required levels has associated problems of taste. Liyanage *et al.* (1997) report trials to evaluate the comparative efficacy of chlorine, chlorine dioxide, sodium thiosulphate, chlorite and chlorate to disinfect drinking water contaminated with *Cryptosporidium*. They conclude that chlorine dioxide is the effective disinfectant which will inactivate the oocysts. Slow sand filtration is an effective treatment but requires large areas of sand-beds at the treatment works. Timms *et al.* (1995) report experimental removal of 99.997% of the oocysts by this method. Advances in filtration technology to bring potable waters to required Environmental Protection Agency standards are exemplified by Algiers and King (1997). They show that filtration can produce required clarity and remove 99.9% of particles in the 4–5 μm and 5–15 μm range, the sizes of *Cryptosporidium* sp. and *Giardia* sp.

Any model estimation of the risk to public health assumes good quantitative data on all the controlling parameters. Marshall (1996) and Teunis *et al.* (1997) conclude that the largest error in current calculations for drinking water risk comes from the uncertainty in the estimate of oocyst removal efficiency at the treatment works. Given the diffuse nature of potential sources of *Cryptosporidium* it becomes clear that defending the river supply system will involve disconnecting areas of overland flow from pastoral agricultural land with some kind of sedimentation buffer. Foster *et al.* (1997) have shown that source areas at risk from colour, *Cryptosporidium* and a variety of other problems can be identified using generic GIS, highlighting areas for intervention (Figure 8.3).

Algal blooms

Algal blooms are a potential hazard in upland reservoirs and canals, although problems are likely to be less acute than in lowland and less acid catchments (NRA, 1990).

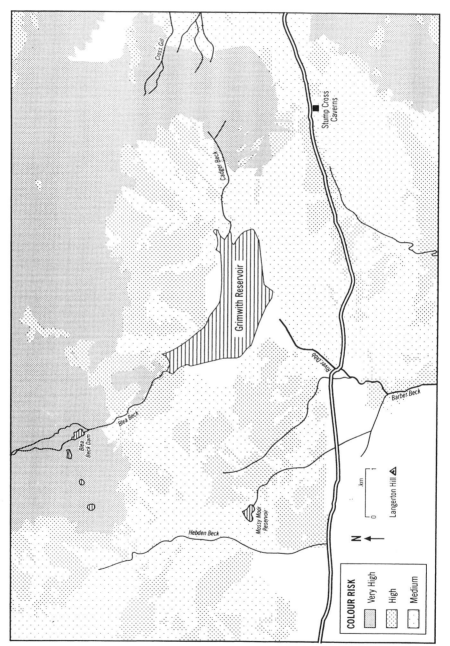

Figure 8.3 GIS identification of areas at risk from colour in runoff

Cyanobacteria warnings are routinely posted by Yorkshire Water at a number of reservoir sites each summer and while these do not deter ramblers around the sites they represent a potential degrading of amenity and recreation value. The problems associated with cyanobacterial blooms and control options have been considered by Howard *et al.* (1996). For an individual lake or reservoir Howard *et al.* (1995) have shown that the growth patterns of cyanobacteria can be modelled in response to biological and environmental factors. The growth of a bloom and its movement across a reservoir in response to wind and wave action can be simulated. However, routine forecasting of this type requires specific site information which would demand a monitoring programme well beyond that currently in place. Kneale and Howard (1997) show that although sampling of cyanobacteria is widespread, covering most reservoirs and lakes in Yorkshire, it is also highly sporadic. Sampling is normally on a snapshot basis and often after a problem has been noted. Statistical analysis of data from 18 sites visited on at least six occasions allows some description of the local position to be made. They are interesting case examples but cannot be used for reliable forecasting.

Coloured runoff

Upland peat soils, rich in humic acids and the products of microbial decomposition, can generate highly coloured runoff. The catchment manager is likely to find colour-enriched waters after drought periods, and where peat catchments are subject to fires, ditching and from sites with serious erosion problems (Mitchell and McDonald, 1991; Reynolds *et al.*, 1996). Organic products build up in the peat profile when water tables are low and microbiological activity high. The next major rainfall period raises the water table and flushes the coloured organics into the water supply. Options for catchment managers may involve forecasting (Naden and McDonald, 1989), identification of areas at risk (Mitchell and McDonald, 1995) and then the physical exclusion of runoff from catchwater and reservoir systems by turning off coloured intakes (Figure 8.3). Where ditching or burning is carried out on peat catchments the manager should be aware of the later problems of coloured runoff and ensure that there are appropriate buffer-strips, sediment traps or settlement ponds to exclude runoff affecting sensitive supplies.

Manganese

Manganese-enriched potable waters are undesirable in giving taste and staining problems. Correlation between stream water manganese and the active peat flush zone was established by Boon *et al.* (1988), although there is not a clear peak in manganese coinciding with runoff in the autumn as is the case with colour and iron. Heal (1995) has shown that the occurrence of manganese in runoff has a seasonal component but it is derived from more than one hydrological and regolith source. Manganese is produced by microbiological processes in peat, and this component is flushed out of the peat when water tables rise in autumn. The larger source of manganese in upland runoff comes with mineral throughflow or groundwater flow. Manganese is a natural mineral weathering product so when groundwater flows are active, manganese that has weathered into solution will appear in runoff. Manganese can therefore appear as a peak in winter stormflows (Heal *et al.*, 1995, 1997).

DISCUSSION

This paper has highlighted a series of issues of concern to water managers in the Pennine uplands – manganese enrichment, *Cryptosporidium*, algal blooms and coloured runoff – while also pointing out the pressures arising from shorter reservoir storage times, increased demand and more fragile catchments than in the past. The public expects clean, reliable, potable water and these expectations are reflected in ever more stringent national and international regulations.

Improvements in water quality through treatment depend upon the characteristics of the treatment works and the characteristics of the raw waters (Table 8.3). A turbid water may be relatively easily treated whereas water with a high true colour requires a much more sophisticated treatment capacity to be available. In principle, improved scientific understanding of the processes may allow the manager to select or reject particular sources, to turn out and exclude problem supplies. Increased appreciation of the by-products and health risks from alternative treatment processes may lead to financially expensive changes to the treatment processes. Options for blending with other waters, or for drawing raw waters entirely from an alternative source, depend upon the engineering connections from the supply sites to the treatment works (Figure 8.4). Identifying potential risk areas through catchment characterisation, integrating and updating information through GIS systems, presents a powerful tool for the manager (Mitchell and McDonald, 1995). Since the processes governing catchment water quality are transferable, a generic catchment management tool is within reach. However, the critical path to implementation may be constrained by ownership, management and risk avoidance rather than scientific understanding (Figure 8.2).

The cost of river water abstraction in the 1980s was about 10 times greater than upland sources, a price differential that pushed managers towards upland exploitation. Managers chose upland sources but consequent rapid drawdown of impoundments

Table 8.3 *Standard treatment methods for surface waters A1, A2 and A3 to potable water (EC Directive 75/440/EC)*

Category	Treatment type
A1	Simple physical treatment and disinfection, e.g. rapid filtration and disinfection
A2	Normal physical treatment, chemical treatment and disinfection, e.g. pre-chlorination, coagulation, flocculation, decantation, filtration, disinfection (final chlorination)
A3	Intensive physical and chemical treatment, extended treatment and disinfection, e.g. chlorination to break-point, coagulation, flocculation, decantation, filtration, adsorption (active carbon), disinfection (ozone, final chlorination).

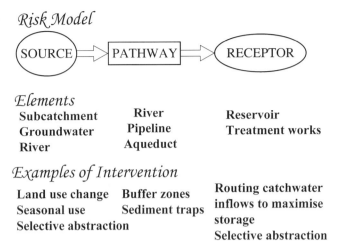

Figure 8.4   *Issues in risk assessment on catchments*

eroded theoretical storage times and while abstracting from the lowest reservoir levels sediment entrainment was a treatment hazard. The balance of abstraction from stressed catchments, especially at times of drought when river water levels are also low, is a water manager's tightrope trick.

We draw the following conclusions:

- Process understanding of the ways in which a number of determinants influence water quality exists already to enable effective potable water management. We note that detection limitations still exist, for example in the case of phenols, and further research will be required when monitoring technologies have improved.
- Similar problems have to be addressed by water managers across the UK, largely because of regulatory requirements; thus a generic approach is required to facilitate the transfer of solutions.
- Digital information for catchment elements allows the automation of such a generic approach.
- It is already possible in a semi-quantitative manner to rank catchments on a fragile–robust profile, thus permitting the identification of catchments at highest risk.
- Such catchments can be accorded a stronger monitoring regime and prioritised for management action, for example by land purchase, establishment of land use management agreements, or can be safeguarded for land disposal.
- Catchments can be characterised to determine their potential to deliver a poor quality raw water.
- Catchment typology can reveal key strategies to improve water quality.

REFERENCES

Algiers, R. and King, K. 1997. Californian utility benefits from 'alternative' filtration systems. *Worldwater and Environmental Engineering*, **20**(7), 12.

Boon, R., Crowther, J. and Kay, D. 1988. *Land use and water quality within the Elan valley*. A Final Report in Completion of Project WR28 for Severn Trent and Welsh Water, Department of Geography, St David's University College.

Casey, T.J. and Chua, K.H. 1997. Aspects of THM formation in drinking-water, *Journal of Water Supply Research and Technology – Aqua*, **46**(1), 31–39.

Environment Agency. 1997. *Local Environment Agency Plan Nidd and Wharfe, Consultation Report*. 165pp.

Fayer, R. and Ungar, B.L.P. 1986. *Cryptosporidium* spp. and cryptosporidiosis. *Microbiological Review*, **50**, 458–483.

Foster, J.A., McDonald, A.T., Macgill, S.M. and Mitchell, I. 1997. GIS-based risk assessment of water supply intakes in the British uplands. In Denzer, R., Swayne, D.A. and Schimak, G. (Eds), *Environmental Software Systems*, Volume 2. IFIP Conference Series, Chapman and Hall, London, 252–258.

Garcia-Villanova, R.J., Garcia, C., Gomez, J.A., Garcia, M.P. and Ardanuy, R. 1997. Formation, evolution and modelling of trihalomethanes in the drinking water of a town, 2. in the distribution system. *Water Research*, **31**(6), 1405–1413.

Heal, K.V. 1995. *Manganese mobilisation and runoff processes in upland catchments*. PhD Thesis, University of Leeds.

Heal, K.V., Kneale, P.E. and McDonald, A.T. 1995.Manganese mobilization and runoff processes in upland catchments. *Proceedings of the Fifth National Hydrology Symposium, BHS*, Edinburgh, 9.11–9.18.

Heal, K.V., Kneale, P.E. and McDonald, A.T. 1997. A hydrochemical basis for managing manganese in water supplies from upland basins. In Webb, B. (Ed.), *Freshwater Contamination*. IAHS Publication 243, 367–374.

Howard, A., Kirkby, M.J., Kneale, P.E. and McDonald, A.T. 1995. GROWSCUM: Modelling the growth of cyanobacteria, *Hydrological Processes*, **9**, 809–820.

Howard, A., McDonald, A.T., Kneale, P.E. and Whitehead, P.G. 1996. Cyanobacterial (blue-green algal) blooms in the UK: A review of the current situation and potential management options. *Progress in Physical Geography*, **20**(1), 53–61.

Hunter, C. and McDonald, A.T. 1991. The occurrence of coliform bacteria in the surface soils of two catchments in the Yorkshire Dales. *Journal of the Institute of Water and Environmental Management*, **5**(5), 534–538.

Kneale, P.E and Howard, A. 1997. Statistical analysis of algal and water quality data. *Hydrobiologia*, **349**, 59–63.

Koivusalo, M., Pukkala, E., Vartiainen, T., Jaakkola, J.J.K. and Hakulinen, T. 1997. Drinking water chlorination and cancer – a historical cohort study in Finland. *Cancer Causes and Control*, **8**(2), 192–200.

Liyanage, L.R.J., Finch, G.R. and Belosevic, M. 1997. Effect of aqueous chlorine and oxy-chlorine compounds on *Cryptosporidium parvum* oocysts, *Environmental Science and Technology*, **31**(7), 1992–1994.

Marshall, J.S. 1996. *Cryptosporidium parvum: detection and distribution in two Yorkshire rivers*. PhD Thesis, University of Leeds.

Marshall, J.S., Kay, D., Kneale, P.E. and McDonald, A.T. 1995. *Cryptosporidium parvum* detection and distribution in the River Wharfe, North Yorkshire. *Proceedings of the Fifth National Hydrology Symposium, BHS*, Edinburgh, 9.25–9.30.

McDonald A.T. 1988. *Bacteriophage tracers and residence times in Nidderdale catchments*. Report to Yorkshire Water.

Mitchell, G. and McDonald, A.T. 1991. Discoloration of water by peat following induced drought and rainfall simulation. *Water Research*, **26**(3), 321–326.

Mitchell, G. and McDonald, A.T. 1995. Catchment characterisation as a tool for upland water quality management. *Journal of Environmental Management*, **44**, 83–95.

Naden, P.S. and McDonald, A.T. 1989. Statistical modelling of water colour in the uplands: the Upper Nidd catchment 1979–87. *Environmental Pollution*, **60**, 141–163.

NRA (National Rivers Authority) 1990. *Toxic blue-green algae*. Water Quality Series No. 2, NRA, London.

Reif, J.S., Hatch, M.C., Bracken, M., Holmes, L.B., Schwetz, B.A. and Singer, P.C. 1996. Reproductive and developmental effects of disinfection by-products in drinking-water. *Environmental Health Perspectives*, **104**(10), 1056–1061.

Reynolds, S.E., Kneale, P.E. and McDonald, A.T. 1996. Quantifying the store of water soluble colour in upland peaty soils: the effect of forestry. *Scottish Forestry*, **50**(1), 22–30.

Teunis, P.F.M., Medema, G.J., Kruidenier, L. and Havelaar, A.H. 1997. Assessment of the risk of infection by Cryptosporidium or Giardia in drinking water from a surface water source. *Water Research*, **31**(6), 1333–1346.

Timms, S., Slade, J.S. and Fricker, C.R. 1995. Removal of *Cryptosporidium* by slow sand filtration. *Water Science Technology*, **31**(5), 81–84.

White, P., Labadz, J.C. and Butcher, D.P. 1996. The management of sediment in reservoired catchments. *Journal of the Chartered Institution of Water and Environmental Management*, **10**(3), 183–189.

Williams, D.T., LeBel, G.L. and Benoit, F.M. 1997. Disinfection by-products in Canadian drinking water. *Chemosphere*, **34**(2), 299–316.

9

Policy, Serendipity and Science: Algal Blooms in Australia

David Ingle Smith

INTRODUCTION

Australians take a perverse delight in describing natural phenomena in their continent as the biggest and best in the world or, if such a claim cannot be universally substantiated, as the biggest and best in the southern hemisphere. Thus, the average rainfall is the lowest for any inhabited continent (Antarctica receives less precipitation) and Australian rainfall and runoff exhibit a variability that is unmatched for any major world region elsewhere, perhaps equalled only in southern Africa. In the early 1990s a new record was claimed: for the longest continuous algal river bloom in the world. This occurred on the Darling River in the summer of 1991/92 and at its peak attained a length of some 1000 km. In contrast to many pollution events, this unenviable claim was exceptionally photogenic and was publicised by the media worldwide. The critical feature of the bloom, dominated by toxic cyanobacteria, was that domestic and stock water normally taken from the river was declared unsuitable for consumption. The ensuing officially declared state of emergency, which lasted for 22 days, required the army to transport special water filtering equipment to numerous small, and often remote, riverine communities. The direct costs of providing the emergency water supply and the implementation of the New South Wales Algal Contingency Plan were A$2 million in New South Wales alone. The indirect costs were much larger, for instance it is reliably estimated that the losses to the regional tourist income alone were some A$9 million (NSWBGATF, 1992). The inevitable result of an event of this magnitude is that governments are motivated to consider measures to mitigate against

Water Quality: Processes and Policy. Edited by Stephen T. Trudgill, Des E. Walling and Bruce W. Webb.

future occurrences. Superficially the government response to the Darling River bloom of 1991/92 represents the usual answer to a catastrophic event, i.e. crisis management. The event also provides an opportunity to analyse the form of such response and especially the links between the environmental science and decision making.

THE BACKGROUND TO ALGAL BLOOMS IN AUSTRALIA

Most of Australia experiences high summer water temperatures which are often accompanied by very low river discharges usually associated with very low velocities. The Darling falls firmly into this category with an exceptionally low gradient, so low that some of its tributaries reverse their flow direction in times of flood! At times of low summer flow, it is more appropriate to consider the Darling as composed as a series of near-stagnant weir pools. Such characteristics are conducive to the formation of algal blooms and these were described well before any post-European settlement with associated changes to the nutrient status of the river network. Charles Sturt, the first European to sight (and taste) the Darling River, commented in the dry season of 1829 that it was too salty to drink. A year later he commented that "... the waters, though sweet, were turbid and had a taste of vegetable decay as well as a light tinge of green". Algal blooms, however, are not restricted to the Darling or to the larger Murray–Darling Basin and, to enlarge on the theme of the biggest and best to the earliest known, Francis (a South Australian government health inspector) contributed the first description in the scientific literature of a toxic bloom which caused the death of cattle (Francis, 1878). There is no doubt that algal blooms are a natural feature of the Australian environment; however, no serious consideration was given to the eutrophication processes, to which they are intimately related, until the early 1970s. Although the database is poor, it is accepted, and is certainly the popular perception, that the magnitude and frequency of eutrophication and algal blooms have increased owing to a variety of developments which range from agricultural practice to sewage disposal. The spectacular Darling event of 1991/92 was not the first major bloom to spark community concern. Two early occurrences prompted government concern that was sufficient to lead to research funding to ascertain the most appropriate management options. These were the blooms and eutrophication of Lake Burley Griffin in Canberra, and the massive accumulation of rotting algae on the beaches of the Peel–Harvey Estuary in Western Australia.

Lake Burley Griffin

The centrepiece of the design of the national capital is Lake Burley Griffin, an artificial lake some 9 km in length and filled in the mid-1960s. The lake soon became affected by unsightly and odorous algal blooms which prompted the funding of research into the source of the nutrients in order to implement acceptable management strategies. In part, the concern was to combat the adverse aesthetic effects on such a prestigious landscape feature situated only tens of metres from the federal Parliament building. The lake was not directly used for water supply but it was a tributary to Lake Burrinjuck, a major irrigation storage located downstream in New South Wales. The algal blooms, and the research they prompted, played a role in the decision to design and construct the Lower Molonglo Water Quality Treatment Plant, a sewerage works which has the capacity to remove virtually all nutrients from its discharge. Since 1983 the Plant has massively reduced the phosphorus content in Burrinjuck reservoir. The Lake Burley Griffin problem, however, was not finally remedied until the upstream sewerage plant at Queanbeyan, also located in New South Wales, was upgraded. A detailed account of the nutrient budgeting and its relevance to management is available in Cullen *et al.* (1978).

The Peel–Harvey Estuary

The algal problem first appeared in the Peel–Harvey Estuary in the late 1960s and deposited rotting algae to a depth of several metres on beaches, traditionally used by the population of Perth for recreation. The Western Australian government funded detailed long-term monitoring and research into the problem which established that there had been a six-fold increase in the phosphorus content of the input river since the commencement of European clearing and farming a hundred years or so before. The solution included the reduction of upstream fertiliser use, which cut phosphorus inputs by an average of 40%, and the construction of the Dawesville Channel, at a cost in excess of A$50 million, which improved the circulation of the algae-infested near-enclosed estuary. These management decisions appear to have been successful and are described and evaluated in Humphries and Robinson (1995).

Both of these early government responses to algal blooms were prompted by the perceived necessity to improve recreational water quality and were not directly related to the toxic effects on water supply. However, in both instances the research was specifically related to a local problem; wider national concern and research into eutrophication and algal problems was accorded a very much lower priority.

GOVERNMENTS, WATER AND THE ALGAL PROBLEM

Any discussion of resources in Australia is complicated as the Constitution allocates land and resource matters to the States. The Commonwealth (federal) government can recommend legislation and management goals and objectives, and assist by funding these, but ultimate responsibility remains with the States. For example, Australia has yet to agree upon a nationwide set of drinking water standards. The situation is further complicated by the role of 900 or so local governments which, especially in the sparsely populated regions, provide water and sewerage services and have considerable power over land use zoning decisions. Multilevel governance is further complicated by the role of the Murray–Darling Basin Commission (MDBC).

This was first established in 1917, as the River Murray Commission (RMC), and remains an unusual Australian institution because its area of jurisdiction incorporated four States and the Australian Capital Territory (ACT); a more detailed account and bibliography is given in Crabb (1991). The need for an inter-basin organisation was recognised at Federation in 1901; the initial problems were those of river navigation (no longer of significance) and water allocation. The governments that constituted the Commission were those of the Commonwealth, New South Wales, Victoria and South Australia; Queensland only joined as a full member in the early 1990s. The ACT (initially represented by the Commonwealth government) is still not a member although the national capital of Canberra is by far the largest urban centre in the Basin. The implementation of decisions of both the earlier RMC and the current MDBC requires full consensus of all the constituent governments.

The total area of the Murray–Darling catchment is close to a million square kilometres, approximately twice the size of France. The population, some 1.8 million, is small but the Basin represents the agricultural heartland of the nation with over 75% of the irrigated area of Australia, half of the sheep and cropland and 25% of the cattle and dairy production. An unusual statistic is that some 28% of the total water of Australia is used in the Basin to irrigate pasture, more than the combined national total for industrial, commercial and residential use! The environmental problems of the catchment are immense, both dryland and wetland salinity are severe (and increasing in area affected) and major dam building since the 1950s has led to over-allocation of water for irrigation. So severe are these problems that the constituent governments of the MDBC agreed to a total "cap" on any new allocation from mid-1997.

Government Reviews of Eutrophication and Water Quality

The first comprehensive review of eutrophication for the inland waters of Australia, by Wood (1975), was jointly funded by the State and Commonwealth governments. The objective was to review research into eutrophication in Australia and worldwide, to assess the applicability of overseas studies for Australia and to identify areas that merited further research. The summary and conclusions (Wood, 1975, p. 185) made it very clear that:

> Long-term investigations on Australian waters which yield meaningful answers on basic factors of fundamental importance to eutrophication and its problems are only of recent times and are far from adequate. It is therefore not possible at this time to make categorical statements as to the applicability of the overseas information to Australian inland waters.

The paucity of data was not restricted to eutrophication but extended across the whole field of water quality and the lack of systematic monitoring was specifically highlighted. Records of changes in eutrophic status were especially sparse with available information based on anecdotal accounts, records of complaints about the taste of drinking water and the occasional newspaper report of stock deaths due to drinking from farm dams or rivers. Only rarely was there any identification of the algae involved. Thus, it was impossible to match possible changes in eutrophic status against base line (or natural) conditions.

The review by Wood contained a long list of recommendations, many designed to improve water quality monitoring and data analysis. There was, however, no noticeable response by governments at any level. A major comprehensive federal government review of the nation's water resources, collectively termed *Water 2000*, was published in the early 1980s. The account of water quality (Garman, 1983) again stressed the parlous state of water quality monitoring and the poor quality of water supplied to small regional centres. Later reviews and studies by State governments re-echoed the same sad story. For instance, a State of Environment Report on the rivers of Victoria (VICOCE, 1988) reported that the data on chemical characteristics for 90% of the State's rivers were insufficient to establish trends or base line conditions. A more recent review for New South Wales (NSWEPA, 1995), after a more comprehensive monitoring system had been introduced, reported that there was "... currently no long-term baseline water quality data". Even more alarming and significant are the conclusions of the *State of the Environment Australia 1996* report that "water quality monitoring and data analysis are poor and deteriorating", that the "situation is very poor nationwide" and that the trends for nutrients are "unclear but the current situation is poor" (SEAC, 1996, p. ES-20). Indeed, each new report paints a situation that is more alarming than the last, to the extent that the process has become akin to national self-flagellation.

Moves Towards Algal and Nutrient Management Strategies

The Darling River bloom of 1991/92 served to highlight these years of neglect. However, it would be over-simplistic to conclude that there had been no prior information gathering on nutrients and eutrophication. One immediate response to the event was to implement a reporting system to record the locations of major blooms. Occurrences of major blooms for the water year 1992/93 are illustrated in Figure 9.1, although this pre-dates a formal and systematic reporting system. The majority of the incidents are within the Murray–Darling catchment and, even in the period prior to the Darling bloom, the MDBC had been to the fore in promoting research and management to control the problem. The reorganisation in the mid-1980s that led to the MDBC heralded a new approach to resource management based on a catchment-wide National Resource Management Strategy. As a precursor to the formulation of the Strategy a major review was initiated, and published as the *Murray–Darling Basin Environmental Resources Study* (MDBMC, 1987). Eutrophication and the problem of algal blooms are stressed throughout and are specifically mentioned as a major adverse feature of water quality for many of the rivers and especially for the trunk streams of the Murray and Darling Rivers.

The study (MDBMC, 1987) contains well over a hundred specific recommendations, many of which emphasise the need for data collection, monitoring, analysis and the development and implementation of common resource management principles, policy and guidelines between the constituent governments. The need for integrated catchment management, involving all three levels of government (federal, state and local), linked to community involvement, was a recurring theme.

The recommendations of the resource study led to the formation of Inter-Government Working Groups and one of these, established in 1990, was specifically charged with developing a "comprehensive nutrient management study for the Murray–Darling Basin". The first task was to review current knowledge and to assess State guidelines and monitoring requirements in relation to nutrients. This was undertaken on behalf of the Commission by Gutteridge, Haskins and Davey. Their report, *An investigation of nutrient pollution in the Murray–Darling River system*, was published in

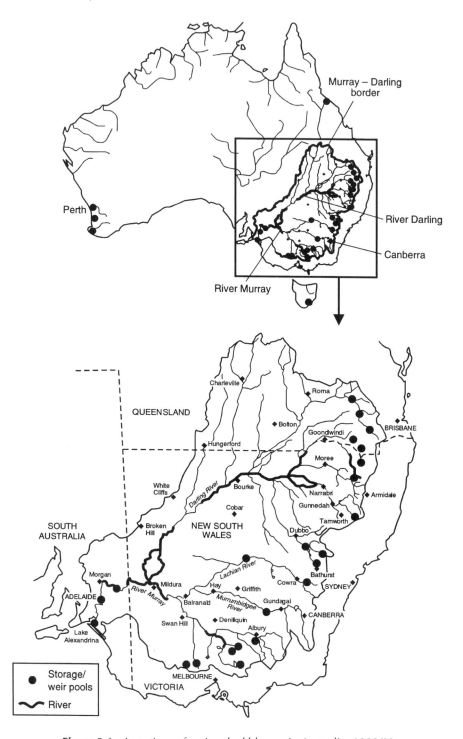

Figure 9.1   *Locations of major algal blooms in Australia, 1992/93*

early 1992 (GHD, 1992). It is significant that the study was in near-final form when the now infamous toxic algal bloom occurred on the River Darling. It would appear that the scientific background and review of the river nutrient problem was at long last starting to catch up with reality.

The report is detailed and represents a major contribution to the study of nutrients and their sources in the Murray–Darling Basin. The budgets assigned values of nitrogen and phosphorus to inputs from differing land uses, both point and non-point, under differing flow conditions. The detail of the nitrogen and phosphorus loadings by source, expressed in terms of dry, average and wet years, is essential to an understanding of the nutrient budget and provides background that is essential in order to assess management options; it is summarised in Figure 9.2. However, caution is required as the data on which the budgets are based have a number of shortcomings. For instance, GHD (1992, p. 4–7) preface the discussion of the inputs from municipal sewage treatment plants with:

> ... in almost all cases there are few or no recorded data available on nutrient concentrations in the effluent from the plants ... in many cases also no information is readily available on annual effluent output.

The problem of the rapid growth in number and size of cattle feed lots and piggeries was stressed but the greatest need was to obtain data on inputs from diffuse nutrient

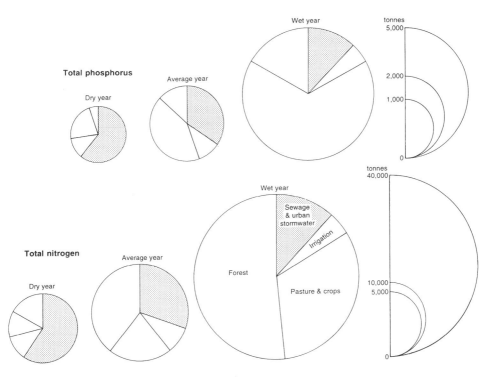

Figure 9.2   *Nutrient budgets for the Murray–Darling Basin for wet, average and dry years. Based on GHD (1992)*

sources, i.e. those associated with non-irrigated agriculture or from forests. In addition to nutrient budgets, State regulations, guidelines and codes of practice were reviewed and, as is so often the case in Australia, demonstrated the lack of comparability and differences in the perceived importance of particular categories of nutrient source between the states.

The preface to the study, by the President of the Murray–Darling Basin Commission, stressed that the next stage was the need for consultation with government agencies, community groups and concerned citizens in order to develop priorities for remedial action. The release of the study in January 1992 could not have come at a more opportune time as it coincided with the state of emergency created by the algal bloom on the Darling River.

Immediately following the release of the nutrient study the MDBC convened 14 technical advisory groups to advise on separate topics related to the management of algae. The topics included algal ecology, nutrient problems posed by different land uses and point discharges, and techniques to treat water for potable supply. The findings of these groups were published by the MDBC to provide input to the development of an overall algal management strategy (MDBC, 1994). This forms the basis for active implementation by the constituent States of the MDBC. Before describing these, it is necessary to outline the broader Australian nation water reforms.

AUSTRALIAN WATER REFORM

There is no doubt that progress with policies to ensure better algal management of Australia's inland waters has gained immeasurably from the impetus provided by the massive Darling River bloom. Notwithstanding this spectacular event, major reforms were, and still are, underway at all levels of government. Milestones in this process have been:

- December 1992. The adoption by the Commonwealth and all State governments of the National Strategy for Ecological Sustainable Development (ESD), the aim of which was to balance economic development with a sustainable environment.
- February 1993. The Council of Australian Governments' (COAG) framework on a Strategic Water Policy.
- August 1993. The Hilmer Committee report on National Competition Policy.
- February 1994. Endorsement by COAG of reforms to the Australian water industry.
- March 1995. Adoption of the National Competition Policy, of which that for water reform is most appropriate to this account.

Collectively these reforms represent a sea change in policy; details are given in a recent account of Australian water resources by Smith (1998). The majority of the water reforms are related to changes in economic and institutional arrangements such as water allocation, consistent water pricing and the introduction of inter-state transferable water entitlements. The significance for algal management is that the reforms are to be undertaken within a setting that fully recognises the importance of ecological sustainable development. The key is that all States and Territories will undertake studies in order to allocate environmental flows for all Australian river catchments; such flows are to be regarded as a legitimate water use.

These policies carry considerable weight, if only because the Commonwealth government in the period to 2002 will make available three tranches of payments to the States. The total funding available to the States, after successful completion of the water reforms, is A$1.2 billion. The policies outlined above have spawned subprogrammes of importance to the algal problem, for instance the establishment, in mid-1995, of the National Eutrophication Management Program. Its goal is to "...undertake the research and communication activities necessary to reduce the frequency and intensity of harmful or undesirable algal blooms in Australian fresh and estuarine waters". The MBDC and the federal Land and Water Resources Research and Development Corporation (LWRRDC) oversee the programme and make available funds to achieve its aims.

Thus, algal and nutrient studies and research are integrated into the new national strategy for water resource management, with their profile heightened because of the events on the Darling in 1991/92. Many of the reforms to the water industry are those associated with economic rationalism and corporatisation, and it is not impossible that these may progress towards a model closer to that of full privatisation. However, environmental consideration associated with rivers and fluvial ecology are now firmly incorporated into the reform process, greatly assisted by the availability of federal funding as a reward for the States and Territories achieving specified goals for policy design and implementation. These moves can be interpreted as the federal government providing leadership to adopt nationwide guidelines for water resource development. The reforms make frequent use of the word "guidelines" rather than "standards"; any firmer wording is traditionally unacceptable to the State governments!

In addition to formal government agreements on policy and the application of market economics, Australia has witnessed a dramatic growth in community awareness and involvement with the biophysical environment. This was initiated by an unlikely accord between the National Farmers Federation and the Australian Conservation Foundation. It was formalised, in July 1989, when Prime Minister Hawke announced the Decade of Landcare and the National Landcare Program with initial funding of A$340 million. The goal was to promote the sustainable management of land, water and vegetation by providing financial assistance to local community groups. The Landcare movement met with immediate success that far surpassed the expectations of even its firmest advocates. Growth in the number of Landcare groups was rapid and it is reliably estimated that they now number over 4000. This has been matched by the formation of other community-based groups with related interests. These include Streamwatch, Drainwatch, Water Watchers, Saltwatch and Watertable Watch; there are also associated groups that promote education programmes for schools, such as Ribbons of Blue. The Landcare programme is now assisted, in all States, by agencies that further the aims of total catchment management.

The future directions of this nationwide expression of community awareness and interest in sustainable development are unclear but as an example of widespread grass roots concern for environmental degradation it must rank as one of the best movements of its kind, if not *the* best, on a world scale. For many communities, algal blooms, as a very visible manifestation of the misuse of inland waters, are of special concern and play a not insignificant role in further prompting governments to improve the state of the nation's inland waters. A detailed account is available in *Landcare: Communities Shaping the Land and Future* by Campbell (1994).

AFTER THE DARLING BLOOM

The Darling River bloom caused individual States to set up task forces and project teams to deal with immediate problems. Their recommendations led to permanent bodies in each State with strategies based upon the overall 1994 Algal Management Strategy of the MDBC. For example, the Blue-Green Algal Task Force in New South Wales was established initially to deal with the day-to-day needs of providing potable water to a large number of remote rural settlements strung like beads along a 1000 km stretch of the Darling River. The affected population was small, only a few thousand, but the problems extended far beyond the provision of measures for traditional water supply plants, to how to provide drinking water for humans and stock for hundreds of individual homesteads. The New South Wales Task Force reported in late 1992 (NSWBGATF, 1992) and was replaced by a permanent State Algal Coordinating Committee.

Similar bodies, although with less urgent problems of emergency management, were established in Victoria, for instance the Victorian Blue-Green Algae Project Team, and in Queensland and South Australia. Although attracting less publicity than the Darling emergency, major water storages in southeast Queensland were also adversely affected by toxic blue-green algal blooms. The period of the early 1990s was one of major, El Niño-linked drought throughout eastern Australia which provided the environmental trigger for widespread blooms.

All of the component States of the MDBC have now adopted a common alert strategy linked to the concentration of toxic blue-green algae. There are two critical concentration levels: domestic users are alerted when the cell count attains 2000/ml and the high alert level is declared at 15 000/ml. At the initial level water can only be used for human consumption and food preparation if regular, at least weekly, toxicity testing is undertaken and there are no toxins, taste or odour. Once the high alert level is attained water cannot be used for any domestic purpose, for stock watering or irrigation and there are severe limitations on use for primary and secondary recreation, including boating and fishing. The alert system is managed by regional committees, of which there are eight in New South Wales. The committees oversee the monitoring, the recognition of algal species, testing for toxins, etc. They also coordinate emergency procedures. The latter include the provision of signs indicating toxicity and the maintenance and distribution of stocks of powdered activated charcoal to maintain a supply of potable water.

Each of the States has a body that oversees algal management; that for New South Wales is selected as an example to illustrate the range of activities. For New South Wales these are described in the biennial report, for 1994–96, of the NSW State Algal Coordination Committee (NSWSACC, 1996). The strategy can be classified into:

- minimising the effects of algal blooms
- managing blooms
- managing the causes of blooms.

1. *Minimising the effects.* There are subsidies for drilling bores to alleviate supplies at times of algal alert; by mid-1996 over a hundred applications had been approved mainly along the main trunk of the Murray and Darling Rivers. Stores of powdered or granulated activated charcoal are strategically located to avoid the costs of storage to smaller councils. The Regional Algal Coordinating Committees have the responsibility for the monitoring and analysis of toxic algae. At the State level, "algal watch kits" have been issued to assist storage operators and others to identify the various species of algae.

2. *Managing the blooms.* This is a component of the broader management of inland water linked to the introduction and implementation of environmental flow strategies within the COAG reforms. For regulated river systems, i.e. those with major dams, a significant improvement is the recognition of the need for flows to assist with the management of blooms, such as the release of water from storages for flushing. Other components are the requirement to review the needs for river weirs and, where appropriate, to remove them. The broader strategy in New South Wales involves the installation of an integrated monitoring system for water quality and flow, and much of this is already in place. The final stage is for workshops and public hearings to formulate the size and timing of environmental river flows; algal management is an integral component of this.

3. *Managing the causes.* A key feature is the production of nutrient management plans for all major rivers within the State; these are currently in final form for about half of the State. Attention has also been given to the management of various components of nutrient generation, ranging from the labelling of detergents to a rolling programme to fund improvements to council sewerage treatment plants. Some councils have participated in pilot schemes to use effluent for the irrigation of tree plantations. The Hunter Water Corporation, one of the larger water agencies in the State, is in the process of establishing an improved treatment plant at a cost of A$300 million. Other, often smaller, councils which are members of the State's Local Government Phosphorus Action Plan aim to increase community awareness in order to limit the use of phosphorus-rich detergents.

The awakening of public and political interest in blue-green algal contamination as a visible and major adverse environmental impact was an additional spur for governments, especially local councils, to seriously consider how to combat algal and nutrient problems.

POLICY AND SCIENCE

The seriousness of algal contamination only became a matter of national concern in the early 1990s. This served to highlight not only the poor Australian record in the field of algology but also the paucity of data on all aspects of water quality. Albeit a response to a major water supply crisis, the last few years have witnessed a remarkable improvement in research into algal ecology and nutrient budgets with the establishment of monitoring both for water quality and the occurrence of blooms. The time scale of the data collected is still insufficient, especially with the extreme variations in Australian rainfall and runoff, for any form of rigorous analysis to be undertaken either for algal occurrence or for nutrient content. This creates problems for management agencies as the algal crisis demanded action but within a context where, by any standards, the background scientific information was very thin. This problem is recognised in the foreword to the algal strategy of the MDBC (1993, p. ii), in which the President of the Commission stated:

> ... it is clear that we cannot sit back and wait for the results of research before taking action ... not all of the [recommended actions] could be implemented with the same degree of confidence, but it is important that we proceed with the most promising "best bet" action making sure we monitor their effectiveness ... this will enable progressive review and refinement of those actions.

In practice, this represented the only realistic approach and many of the actions recommended in the strategy will undoubtedly serve to reduce the problem of contaminated water supplies and to improve the overall health of Australia's inland waters. However, many of the options involve considerable financial outlays. Notable among these are modifications to water storages to ensure that domestic supplies are free from toxins and measures that could perhaps "seal off" the supply of nutrients stored in bottom sediments. As Australian research into algal problems becomes more sophisticated it is becoming apparent that all aspects of algology and nutrient supply are complex. A particularly useful background to such problems is given in *Managing Algal Blooms* (Davis, 1997). This presents the results of nine detailed scientific studies undertaken by various divisions of the Commonwealth Scientific and Industrial Research Organization (CSIRO) during the period 1992–95, i.e. research funded directly as a response to the algal crisis of the early 1990s. The studies are spread across the Australian continent and vary in research topic. It is not appropriate to precis all of the results here, but it is pertinent to highlight some of the findings, especially those that question some of the accepted "principles" and those which question the applicability of relying on overseas results.

1. Two of the studies (Thompson *et al.*, 1997; Douglas *et al.*, 1997) were based on the Swan River, on which Perth is situated, and found that nitrogen rather than phosphorus was the limiting factor, especially during the major bloom summer season. Another surprising feature was that the increases of phosphorus to the water column from the bottom sediments did not increase when the lower layers became oxygen deficient. The Swan River is relatively free of cyanobacteria despite the problems encountered in the Peel–Harvey Estuary which is located on the coast only some 50 km south of Perth. The studies also raise the question that suppression of one form of algae may lead to the increase in other species; for the Swan River this could mean the spread of toxic cyanobacteria.
2. Studies in inland New South Wales (Webster *et al.*, 1997; Chambers *et al.*, 1997) highlight the need to break down the stratification that is common in summer in the long shallow weir pools that form part of the irrigation system used in many

rivers in the Murray–Darling Basin. The stratification favours the growth of cyanobacteria which have the buoyancy necessary to exploit the light that is only available in the surface layers; Australian rivers and lakes have a remarkably high turbidity even in summer. Possible remedial measures include flushing the weirs at regular intervals at times of low flow to break up the stratification, although this would reduce the water available for irrigation, or to use siphons and other release mechanisms to break up the thermal stratification.

3. A study of water storages in southeastern Queensland by Jones (1997) questions whether improvements to catchment management in a region with intense tropical rainfall is a viable strategy to reduce nutrient inputs. The key is again to introduce measures to break up the thermal stratification. Useful information is also given as to the frequency and style of monitoring necessary to employ the algal alert system that is part of the State and national strategy.

4. Of special note is the contribution by Blackburn *et al.* (1997) which considers the question of why cyanobacteria blooms are toxic. The significant findings are that the genetic basis of bloom-forming organisms is the key to toxicity and that toxicity changes with the age of the bloom. Perhaps most significant of all is that Australian species, even when morphologically identical to non-Australian species, are genetically different and have different toxicological properties. Tentatively this raises the possibility of a biological solution by the use of bacteria to provide enzymes that promote the breakdown of the toxins.

5. Cuddy *et al.* (1997) developed and trialled a computer-based catchment management support system which allows local groups to use data on land use and soils etc., to predict the impact and costs of changes in management to nutrient supply.

The studies outlined above are matched by other work. For instance Martin and McCulloch (1995) have demonstrated, using environmental isotope techniques, that much of the phosphorus in sediments in the Namoi catchment in northern New South Wales is derived from the lower soil horizons typically exposed in gully erosion or stream banks, rather than for the phosphorus in fertiliser runoff. The message is that the management options to combat algal problems require a continuation of specialised and inter-disciplinary research in order to find optimal solutions. The adoption of a common solution from one part of Australia to another is not valid and there are even greater problems in assuming that Australian species react in a similar manner to those from overseas. However, it must be added that the volume and detail of the research undertaken in Australia since the Darling algal disaster is impressive. Pre-1990 it was uncommon for the species responsible for the blooms to be even noted, now the question has moved to the genetics of individual species at specific locations!

CONCLUSION

The occurrence of the world's largest algal bloom on the Darling River in 1991/92 was sufficient to attract international media coverage and was instrumental in the selection of the topic to illustrate the interaction of policy, management and science. The event not only caused a national emergency, it also provided the stimulus for a better understanding of the causes in order to mitigate against their recurrence. Earlier algal disasters, in Western Australia and in Canberra, had promoted locally relevant response and mitigation but the potential to extend these elsewhere was not realised. The early years of the 1990s also coincided with a broader acceptance of the shortcomings of Australian water quality research and policy. The Darling incident contributed to the urgency to implement water resources policy reforms and the algal aspects have gained from the broader realisation of the need for healthy rivers.

The few years since the Darling incident have witnessed a dramatic upsurge in scientific research into all aspect of the algal problem together with implementation of often expensive mitigation solutions. Traditionally the scientific community is reluctant to recommend management solutions until the science is fully understood; all too often governments have been reluctant to fund research until there is a crisis. The latter was undoubtedly true with a disregard of advice for over 20 years of the recommendations of government-funded reviews to improve water quality monitoring in Australia. However, the Darling algal crisis, especially when it threatened the supply of potable water, led to a loosening of purse strings for research and was met by the willingness of scientists to recommend options based on a less than ideal database. As the science base for algal management improves it is becoming apparent that the comments of Trudgill and Richards (1997) that the "...contexts of scale, place and time are highly relevant in tackling environmental issues" are very true.

The current needs for Australian studies of algal management are for a better understanding of process, at the local level and under conditions of extreme rainfall and runoff variability, before management options can be presented with an acceptable degree of certainty. It is much too early to evaluate the success, either in scientific or cost–benefit terms, of the management options that have been introduced so far. However, all can be regarded as acceptable in terms of the precautionary principle. The degree to which this Australian case study is relevant elsewhere is less certain. This reflects the serendipity of the relationship between science and policy which in Australia requires the interaction of the Commonwealth and eight State and Territory governments. Perhaps the most laudable outcome of the whole process has been a very real increase in concern at community level.

REFERENCES

Blackburn, S., Bolch, C., Jones, G., Negri, A. and Orr, P. 1997. Cyanobacterial blooms: why are they toxic? The regulation of toxin production and persistence. In Davis, J.R. (Ed.), *Managing Algal Blooms*. CSIRO Land and Water, Canberra, 67–78.

Campbell, A. 1994. *Landcare: Community Shaping the Land and Future*. Allen and Unwin, Sydney.

Chambers, L., Olley, J., Crockford, H. and Murray, A. 1997. Conditions affecting the availability and release of phosphorus from sediments in the Maude Weir Pool on the Murrumbidgee River, New South Wales. In Davis, J.R. (Ed.), *Managing Algal Blooms*. CSIRO Land and Water, Canberra, 41–50.

Crabb. P. 1991. Resolving conflict in the Murray-Darling Basin. In Handmer, J.W., Dorcey, A.H.J. and Smith, D.I. (Eds), *Negotiating Water: conflict resolution in Australia*. Centre for Resource and Environmental Studies, Australian National University, Canberra, 147–159.

Cuddy, S., Young, B., Davis, J. and Farley, T. 1997. Trialing the catchment management support system in the Murrumbidgee catchment; New South Wales. In Davis, J.R. (Ed.), *Managing Algal Blooms*. CSIRO Land and Water, Canberra, 103–113.

Cullen, P.W., Rosich, R.S. and Bek, P. 1978. *A phosphorus budget for Lake Burley Griffin and management implications for urban lakes*. Technical Paper No. 31, Australian Water Resources Council, Canberra.

Davis, J.R. (Ed.) 1997. *Managing Algal Blooms: outcomes from CSIRO's multi-divisional blue-green algal program*. CSIRO Land and Water, Canberra.

Douglas, G., Hamilton, D., Gerritse, R., Adeney, J. and Coad, D. 1997. Sediment geochemistry, nutrient fluxes and water quality in the Swan River estuary, Western Australia. In Davis, J.R. (Ed.) *Managing Algal Blooms*. CSIRO Land and Water, Canberra, 15–30.

Francis, G. 1878. Poisonous Australian lake. *Nature*, **18**, 11–12.

Garman, D.E.J. 1983. *Water Quality Issues*. Water 2000, Consultant's Report no. 7, Department of Resources and Energy, AGPS, Canberra.

GHD. 1992. *An investigation of nutrient pollution in the Murray–Darling River system*. Murray–Darling Basin Commission, Canberra.

Humphries, R. and Robinson, S.J. 1995. Assessment of the success of the Peel-Harvey system management strategy – a Western Australian attempt at integrated catchment management. *Water Science Technology*, **32**(5–6), 255–264.

Jones, G. 1997. Limnological study of cyanobacterial growth in three south-east Queensland reservoirs. In Davis, J.R. (Ed.), *Managing Algal Blooms*. CSIRO Land and Water, Canberra, 51–66.

Martin, C.E. and McCullough, M.T. 1995. Sources of phosphorus in rivers of the Namoi Basin, New South Wales: Nd-Sr isotopes and rare-earth elements constraints. In *MODSIM Conference Proceedings*, Newcastle, 27–30 November 1995.

MDBC. 1993. *Algal Management Strategy – technical advisory group report*. Murray–Darling Basin Commission, Canberra.

MDBC. 1994. *The Algal Management Strategy for the Murray–Darling Basin*. Murray–Darling Basin Commission, Canberra.

MDBMC. 1987. *Murray–Darling Basin Environmental Resources Study*. Murray–Darling Basin Ministerial Council, Canberra.

NSWBGATF. 1992. *Blue-Green Algae: a Summary of the Final Report*. Blue-green Algae Task Force, New South Wales, Department of Water Resources, Parramatta.

NSWEPA. 1995. *New South Wales State of the Environment Report*. New South Wales Environment Protection Authority, Sydney.

NSWSACC. 1996. *Implementing the New South Wales algal management strategy – biennial report 1994–96*. New South Wales State Algal Co-ordinating Committee, Department of Land and Water Conservation, Parramatta.

SEAC. 1996. *Australia State of the Environment 1996*. CSIRO, Melbourne.

Smith, D.I. 1998. *Water in Australia: resources and management*. Oxford University Press, Sydney.

Thompson, P., Adeney, J. and Gerritse, R. 1997. Phytoplankton in the Swan River, Western Australia: research results 1993 to 1996 and management implications. In Davis, J.R. (Ed.), *Managing Algal Blooms*. CSIRO Land and Water, Canberra, 1–14.

Trudgill, S.T. and Richards, K. 1977. Environmental science and policy: generalizations and context. *Transactions of the Institute of British Geographers, New Series*, **22**(1), 5–12.

VICOCE. 1988. *Victoria's Inland Waters: state of the environment report 1988*. Office of the Commissioner for the Environment, Melbourne.

Webster, I., Jones, G., Oliver, R., Bormans, M. and Sherman, B. 1997. Control strategies for cyanobacterial blooms in weir pools. In Davis, J.R. (Ed.), *Managing Algal Blooms*. CSIRO Land and Water, Canberra, 31–40.

Wood, G. 1975. *An Assessment of Eutrophication in Australian Inland Waters*. Technical Paper no. 15, Australian Water Resources Council, Canberra.

10

Towards Groundwater Protection in the UK: Problems of Integrating Science and Policy

C. Soulsby, M. Chen and R. Malcolm

INTRODUCTION

Over 98% of available global freshwater is stored as groundwater in the saturated zone within pores and fractures in rocks beneath the Earth's surface (Hiscock, 1994; Jones, 1997). Consequently, groundwater constitutes a significant proportion of water resources in most countries and is an important component of the natural environment (IHP, 1991). For example, groundwater accounts for around 35% of public water supplies in England and Wales (NRA, 1993). However, where surface water is limited, particularly in the dry south and east of England, it is by far the dominant source of water and contributes to domestic, agricultural and industrial consumption (NRA, 1993). Groundwater often provides a water supply that is more reliable in quantity and more stable in quality than surface water and thus has economic and operational advantages due to reduced treatment requirements (Robins, 1990). From an environmental perspective, groundwater discharges can sustain important wetland habitats of high conservation value (Hughes and Heathwaite, 1995). Furthermore, groundwater discharges into rivers and streams maintain baseflows and moderate the chemical and thermal regimes (Soulsby *et al.*, 1997), thus exerting a strong influence on aquatic communities (Neal *et al.*, 1997; Soulsby *et al.*, 1998).

Despite these fundamentally important roles, the protection of groundwater has not received the concerns and priorities that have been given to the protection of surface

Water Quality: *Processes and Policy*. Edited by Stephen T. Trudgill, Des E. Walling and Bruce W. Webb.
© 1999 John Wiley & Sons Ltd.

waters (Mackay, 1991). This reflects the fact that groundwater lies below extensive areas of the land surface and is not a visible resource; therefore, in contrast to surface waters, the effects of any pollution are not always immediately apparent (IHP, 1991). Consequently, less attention has been paid to groundwater monitoring and this has dictated that any deterioration in groundwater quality is usually detected at quite a late stage, often when drinking water supplies become threatened or contaminated.

With increased understanding and a recognition of this neglect, policy development on groundwater protection has increased dramatically over the last two decades in most developed countries (Skinner, 1991; Hiscock, 1994). Like many aspects of environmental management, groundwater protection has been an area where scientific understanding has underpinned the development and formulation of policy (O'Riordan, 1994). However, policies inevitably need to be relatively simple and realistic, in addition to being politically and economically acceptable, so that they can be applied. Often this sits uneasily with environmental science which tends to be complex, with a high degree of uncertainty (Jones, 1997). These problems become particularly apparent when considering complex issues of groundwater protection where compromise and pragmatism are required (Adams and Foster, 1992).

This paper examines groundwater protection policies and strategies in the UK and considers how scientific understanding of hydrogeological processes has provided the foundation for policy development and implementation (BGS, 1991; Chen and Soulsby, 1997a). Although the concept of protection involves both the quantity and quality of groundwater resources (Chen and Soulsby, 1997b), this paper will focus on the protection of groundwater quality from pollution. The paper firstly examines the main pollution risks that groundwater is exposed to. The groundwater resources of the UK will then be considered, with the emphasis throughout being on England, Wales and Scotland. The different institutional responsibilities for groundwater management and protection will be described. The scientific principles upon which resource protection strategies are based will then be examined with reference to the concept of groundwater vulnerability to pollution, a concept which is based on the geographical classification of extensive areas of land in relation to aquifer and soil characteristics. At a more detailed scale, source protection strategies will consider the safeguarding of individual groundwater sources by zoning localised land surfaces on the basis of groundwater flow rates. The principles underlying the development of source protection zone strategies are described and the different procedures that can be utilised to define them are reviewed. A case study of the largest groundwater abstraction scheme in Scotland, in the alluvial aquifer of the River Spey, will be presented. This will demonstrate the difficulties of integrating science with aquifer protection strategies that can be implemented and enforced. The strengths and weaknesses of current policies will be examined and the pressing need for internationally defensible and achievable protection programmes will be stressed.

GROUNDWATER AND POLLUTION RISKS

Historically groundwater has not been perceived by environmental managers as having the same priority for protection from pollution as surface waters. This is mainly due to it being beneath the Earth's surface and poorly monitored; hence pollution problems are seldom immediately obvious. Nevertheless, as Figure 10.1 shows, the continuity of the hydrological cycle dictates that any number of activities occurring at the Earth's surface have the potential to introduce pollutants into hydrological flow paths and eventually contaminate groundwater resources (Adams and Foster, 1992; Soulsby, 1996). As such pollution occurs at an early stage in the terrestrial phase of the hydrological cycle, and aquifers can cover extensive areas, it follows that any problems may affect significant areas of groundwater before they become detected.

Some of the major potential sources of groundwater pollution are listed in Table 10.1, together with the nature of the contaminants that they introduce into the hydrological cycle and their relative significance. This, along with Figure 10.1, shows that both "point" and "diffuse" sources of pollution may occur. The former are restricted to small spatial areas, such as septic tanks, whilst the latter are much more widespread such as fertiliser applications on to agricultural land. Moreover some pollution sources may be intermittent (such as accidental spillage) whilst others may be continuous (such as a continually leaking septic tank).

The urban environment poses a number of significant risks to the quality of underlying aquifers in the UK. Sewerage systems can cause problems, as many trunk sewers in old industrial towns were constructed in the nineteenth century and deterioration in the physical structure of these features can lead to leakage of raw sewage which can infiltrate to groundwater (D'Arras and Suzanne, 1993). Organic and bacteriological pollutants can be released from domestic sewage and inorganic chemicals are often also present where industries discharge trade effluents into sewer systems (Lewis *et al.*, 1982). Inappropriately sited discharges to groundwater, such as soakaways from septic tanks, may also cause problems. One of the main concerns comes from the issue of waste disposal, particularly landfill sites where the surrounding geology is permeable and leachate from these point sources has been introduced into groundwater flow paths historically (Figure 10.1). The wide range of pollutants present in waste materials dictates that this can be a particularly serious source of organic material and inorganic chemicals, including trace metals (Robinson and Gronow, 1992). Where industrial sites are present, other risks, often of accidental spillages and leakages, can occur. Former industrial sites can pose more serious problems as they may be characterised by the presence of large areas of contaminated land where chemicals used in former industrial processes pollute drainage water which flows into the underlying aquifer (Lerner *et al.*, 1993; Henton and Young, 1993; Chen *et al.*, 1997b). Petroleum from accidental spillages at industrial sites and petrol stations can also be a serious pollutant that can be mobilised in urban runoff.

Agricultural waste is often recycled on the farm as liquid waste and sludges are often applied over large areas as fertilisers (Foster and Young, 1981). Although primarily composed of organic pollutants, they can have a high bacteriological load which has the potential to be a significant source of diffuse biological pollution. Together with leaching of pesticides and inorganic fertilisers, most notably nitrate, these can be important diffuse pollution sources over large areas of agricultural land (Foster *et al.*, 1986). The storage of farm wastes and agrochemical also has the risk of accidental spillage (MAFF, 1991).

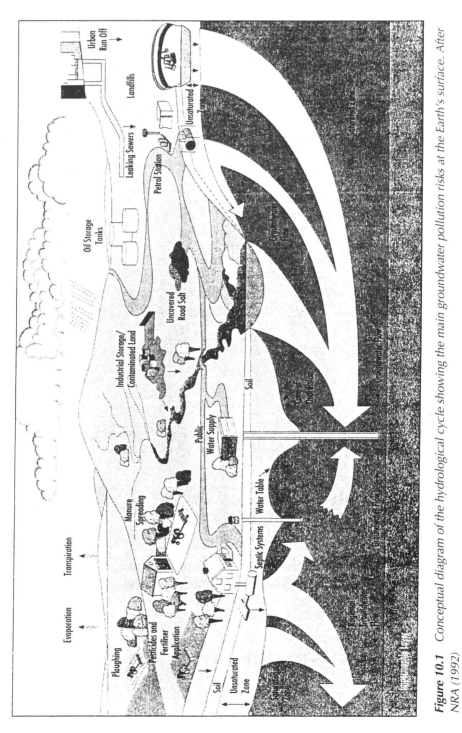

Figure 10.1 *Conceptual diagram of the hydrological cycle showing the main groundwater pollution risks at the Earth's surface. After NRA (1992)*

Table 10.1 *Summary of main sources of groundwater pollution, types of contaminants and their relative significance. After Todd et al. (1976) and Mackay (1991)*

Source	Physical	Inorganic chemical	Trace elements	Organic chemical	Bacteriological
Municipal					
Sewage leakage	Minor	Primary	Secondary	Primary	Primary
Sewage effluent	Minor	Primary	Secondary	Primary	Primary
Sewage sludge	Minor	Primary	Primary	Primary	Primary
Urban runoff	Minor	Secondary	Variable	Primary	Minor
Waste disposal	Minor	Primary	Primary	Primary	Secondary
Septic tanks and cesspools	Minor	Primary	Minor	Secondary	Primary
Agricultural					
Leached salts	Minor	Primary	Minor	Minor	Minor
Fertilizers	Minor	Primary	Secondary	Secondary	Minor
Pesticides	Minor	Minor	Minor	Primary	Minor
Animal waste	Minor	Primary	Minor	Secondary	Primary
Industrial					
Cooling water	Primary	Minor	Primary	Minor	Minor
Process waters	Variable	Primary	Primary	Variable	Minor
Water treatment effluent	Minor	Primary	Secondary	Minor	Minor
Hydrocarbons	Secondary	Secondary	Secondary	Primary	Minor
Tank and pipeline leakage	Variable	Variable	Variable	Variable	Minor
Oilfield waste					
Brines	Primary	Primary	Primary	Minor	Minor
Hydrocarbons	Secondary	Secondary	Secondary	Primary	Minor
Mining	Minor	Primary	Variable	Minor	Minor
Miscellaneous					
Surface water	Variable	Variable	Variable	Variable	Variable
Saline intrusion	Primary	Primary	Primary	Minor	Minor
Transport	Minor	Minor	Minor	Primary	Variable

Any pollution of groundwater is problematic owing to the much longer residence times that water in the groundwater phase of the hydrological cycle generally has compared to surface waters (Jones, 1997). Consequently, whereas rivers may take days or weeks to recover from the chemical effects of a pollution incident, the turnover period for groundwater recovery may be centuries or millennia (Falkenmark, 1990, 1989). Technological solutions for groundwater remediation are invariably expensive and often difficult to implement; in many instances the abandonment of a particular groundwater source is the short-term solution (Kovar and Krasney, 1995). However, while finding alternative water supplies may solve the resource problems created by groundwater pollution, it does nothing to resolve the issue of reduced environmental quality and likely long-term ecological impacts (Hiscock, 1994).

THE UK CONTEXT

Although the UK is relatively small, the geological diversity is large and the country is underlain by aquifers, mainly in sedimentary rocks, which cover extensive areas (Figure 10.2). In England and Wales, the Cretaceous chalk deposits, together with Jurassic and Magnesian limestones, form the major regional aquifers which run in an arc from Dorset in the south to Durham in the north. In addition, Permo-Triassic sandstones form important aquifers in Devon, Cheshire, Shropshire, Lancashire, Cumbria and the Vale of York. In Scotland the hard crystalline basement rocks which dominate the Highlands are relatively impermeable (Figure 10.2a). However, sedimentary rocks from the Carboniferous in the central valley, Permian deposits in the southwest and Devonian rocks in Moray, Caithness and the area immediately south of the Highland Boundary Fault, all act as important aquifers (Robins, 1988). In addition to these main aquifers, rocks with relatively poor hydraulic properties and various kinds of drift deposits can be locally important water resources throughout the UK (Robins, 1990). Extensive aquifers also cover large parts of Northern Ireland (BGS, 1994). Groundwater resources have a long history of exploitation in the UK and the long industrial legacy and high population densities dictate that the risk to groundwater integrity continue to increase.

POLICY DEVELOPMENTS IN GROUNDWATER PROTECTION

Many of the existing problems of groundwater pollution have been the inadvertent consequence of ignorance or poor water resource planning. Nevertheless, in many areas a formidable historic legacy of pollution risk is present which needs proactive management to protect groundwater for existing and future uses (Skinner, 1991). Moreover, proactive policies are required to guide planners, developers, industry and others to ensure that groundwater resources are maintained in the future by appropriate control and management strategies (BGS, 1991).

The last decade has seen a significant increase in efforts to protect groundwater resources in the UK. This has largely paralleled developments in other countries in Europe (Arnold and Willems, 1996; Hyde *et al.*, 1994; IHP, 1991) and North America (Bourne *et al.*, 1995; USEPA, 1987). The current principal competent authorities with legislative powers for pollution control are the Environment Agency (EA) in England and Wales, the Scottish Environment Protection Agency (SEPA) in Scotland and the Environment Service in Northern Ireland. Both the EA and SEPA were created in 1996 as a result of the Environment Act (1995). Both organisations have largely inherited the powers given to their predecessors: the National Rivers Authority (NRA) in England and Wales and River Purification Boards (RPBs) in Scotland. These previous agencies first published strategic statements on groundwater protection, with the NRA publishing its *Policy and Practice for the Protection of Groundwater* in 1992 (NRA, 1992) and the RPBs under the auspices of their national co-ordinating body, the Association of Directors and River Inspectors of Scotland, publishing their *Groundwater Protection Strategy for Scotland* in 1995 (ADRIS, 1995). The Environment Act

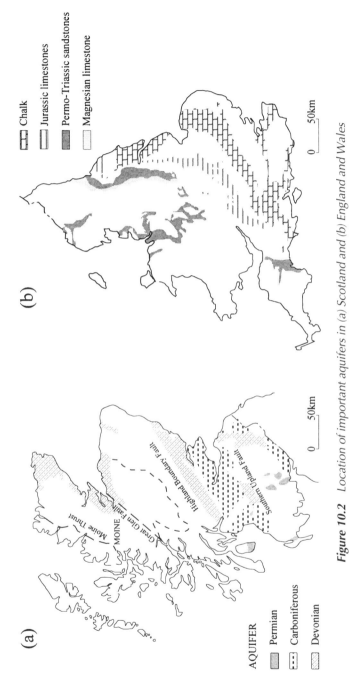

Figure 10.2 *Location of important aquifers in (a) Scotland and (b) England and Wales*

(1995) gave EA and SEPA further responsibilities including the assimilation of some of the waste management functions of local authorities and thus theoretically strengthened their ability to protect groundwater.

These very recent policy developments represent the culmination of a somewhat fragmented approach to groundwater protection in the UK during the preceding years (Skinner, 1991). Powers to protect groundwater from a range of polluting activities stem from a complex series of legislation (see Howarth (1990) for a full review). Of major importance was the 1974 Control of Pollution Act Part II (amended by the Water Act 1991) which included groundwater as controlled water and allowed some degree of protection (Table 10.2). Although a series of powers derived from national legislation (listed in Table 10.2) gave mechanisms whereby competent authorities could control groundwater pollution in the UK, it is increasingly pressure from the European Union (EU) that has forced member states to develop more co-ordinated approaches to aquifer protection (Sheils, 1993). This pressure has come in the form of legally binding Directives such as the Groundwater and Nitrate Directives, and more

Table 10.2 *Selected important legislation relating to groundwater protection in the UK and some direct or indirect roles of EA and SEPA*

Origin	Legislation	Role of competent authority
European	EC Groundwater Directive (80/68/EEC)	Implement directive
	EC Directive to protect water (91/676/EEC) against pollution by nitrates from agricultural sources	To make recommendations on identification of polluted waters and associated vulnerable zones
National	Control of Pollution Act 1974 Part II, amended by the Water Act 1989 and the Water Resources Act 1991	Definition of Water Quality Objectives for controlled water
		Powers to control discharges to groundwaters
		Powers to consent prescribed activities in designated areas
		Powers under regulations to require pollution prevention measures to be undertaken
		Advises on Nitrate Sensitive Areas
		Powers to take remedial action to prevent pollution occurring or continuing
		Designate source protection zones
	Environmental Protection Act (1990)	Powers to control releases from prescribed processes
	National Heritage (Scotland) Act (1991)	Powers to control by licence abstractions for irrigation for commercial agriculture and horticulture
	Environment Act (1995)	SEPA and EA created, further powers for groundwater protection

recently in the form of the Hague Declaration of 1991. This marked a major milestone for groundwater protection in the EU by initiating a draft "Action Programme for Integrated Groundwater Protection and Management" through the European Commission (Arnold and Willem, 1996).

Despite these responsibilities, the EA and SEPA are not the only agencies that have powers relating to activities that are relevant to groundwater management. For example both EA and SEPA may be non-statutory consultees in many planning applications that have the potential to impact upon groundwater resources. Moreover, the diffuse pollution of many aquifers by agricultural activities such as fertiliser applications can in many cases only be prevented by education, persuasion and the voluntary adherence to Codes of Practice (MAFF, 1991). Thus it has been argued that the lead agencies (EA and SEPA) have responsibilities without power, in that whilst certain discharges to groundwatr may be controlled and groundwater quality objectives may be set, the ability to achieve groundwater protection objectives is constrained by lack of statutory powers (Skinner, 1991). The Association of Directors and River Inspectors for Scotland was certainly conscious of calling their groundwater protection document (ADRIS, 1995) a "strategy" rather than a "policy" as it had more limited powers to implement a policy than the EA in England and Wales.

Notwithstanding these institutional constraints, the realisation of groundwater protection objectives needs to be firmly based on a good scientific understanding of groundwater behaviour in relation to the threats that a particular resource is exposed to. Although the scientific processes of groundwater movement and hydrogeochemical interactions in the hydrological cycle are reasonably well understood (Appelo and Postma, 1993), integrating scientific understanding with management approaches is often difficult. The extensive nature of potentially polluting activities, together with the varying effects that these can have on different aquifer types under variable pollutant loadings, creates practical problems for the setting of achievable groundwater protection strategies (Skinner, 1991). In common with most other countries, the strategies developed in the UK are concerned with protection at two scales (Adams and Foster, 1992). At the strategic level, resource protection of extensive aquifers requires an understanding of the vulnerability of groundwater to pollution. At the more localised operational level, management is concerned with protecting individual groundwater sources. These two issues are examined in the following sections.

RESOURCE PROTECTION: CONCEPTS OF GROUNDWATER VULNERABILITY

At the strategic level of groundwater protection, the vulnerability of aquifers to pollution depends upon the interaction of several site-specific factors. The characteristics and depth of soil and unsaturated strata that lie between the land surface and the water table are important, together with the properties of the aquifer itself (BGS, 1991, 1994; NRA, 1992; ADRIS, 1995). The interaction of these factors determines the nature and velocity of the pathway that will route pollutants to the water table via the processes of advection and dispersion (Domenico and Schwartz, 1990). In addition, they will control the degree of attenuation that might restrict the penetration of pollutants through the soil and unsaturated zone. The soil zone in particular is a region of intense biological activity and chemical reactivity, where pollutants can be intercepted. Pathogenic bacteria and viruses from pollution sources can often be eliminated, inorganic pollutants are likely to be involved in precipitation, cation exchange and sorption processes (Soulsby 1996). The extent to which these processes can attenuate pollution migration depends upon the characteristics of the particular pollutant, the degree of hydraulic loading and the specific biogeochemical environment (Adams and Foster, 1992).

At a very simple conceptual level, the way in which these factors interact to determine vulnerability is shown in Figure 10.3. This shows two different situations illustrating the extreme differences between groundwater with high and low vulnerability, respectively. The former is characterised by a thin soil layer, permeable strata and a shallow water table; the later exhibits a thicker soil layer, less permeable strata and a greater depth to the water table. To a certain extent, however, the concept of vulnerability is strongly dependent upon the pollutant in question. Indeed it has been suggested that groundwater protection strategies should be based on individual pollutants or groups of pollutants which exhibit differing potential for mobilisation and transport (Andersen and Gosk, 1987). Whilst such an approach is certainly scientifically defensible, it is difficult to envisage how such strategies could be operationalised as they would be very unwieldy and difficult to implement (Adams and Foster, 1991). A consensus has emerged that a pragmatic approach needs to be taken in representing groundwater vulnerability which, although scientifically based, needs to be relatively simple and generic in order to be utilised in groundwater protection (Mackay, 1991; IHP, 1991).

In 1992, the National Rivers Authority of England and Wales, in conjunction with the British Geological Survey and the Soil Survey and Land Research Centre, published a national groundwater vulnerability classification based on the aquifer properties and soil characteristics (NRA, 1992). The basic features of the classification and rationale behind it are shown in Table 10.3. The soil zone represents the first line of defence against pollutant transport into groundwater. As part of vulnerability mapping, the soils of the UK have been classified into soils of high, moderate and low leaching potential (Table 10.3). The leaching potential is largely determined by the permeability of the soil and the likelihood of biological and chemical interactions allowing attenuation. The characteristics of the aquifer formation then determine whether or not it is highly, moderately or weakly permeable as vulnerability will clearly increase with permeability. Given this vulnerability of permeable aquifers, the three-fold soil classification is used to subdivide highly permeable, or high risk, aquifers. A further important factor is the extent to which any superficial drift deposits are overlying the aquifer and may allow further protection by creating confining conditions or allowing further attenuation.

Vulnerability maps based on this classification scheme have been produced for England and Wales (NRA, 1992), Scotland (ADRIS, 1995) and Northern Ireland (BGS, 1994). Figure 10.4 demonstrates the differing vulnerability of different types of aquifers in northeast England where the major regional aquifer of Magnesian limestone in County Durham runs from north of the Tees to Tynemouth. In places the

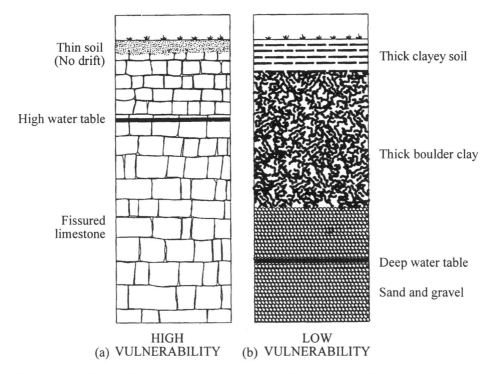

Thin soil (No drift)

High water table

Fissured limestone

Thick clayey soil

Thick boulder clay

Deep water table

Sand and gravel

HIGH
(a) VULNERABILITY

LOW
(b) VULNERABILITY

Figure 10.3 *Vulnerability of groundwater under (a) high risk and (b) low risk environments. After NRA (1992)*

limestone aquifer is highly vulnerable to pollution as a result of thin rendzina soil cover. Elsewhere, however, drift deposits and thicker soil cover render groundwater less vulnerable. The moderately permeable aquifer of Triassic sandstone in the Vale of York, south of the River Tees, is also shown on the map. The Carboniferous uplands of the Pennines and Northumberland consist primarily of minor aquifers of low permeability. In contrast, Jurassic clays south of Teesside and Redcar form non-aquifers.

The purpose of these maps is to provide lead agencies like EA and SEPA with a powerful tool that can be used to influence planners, industry, agriculturists and developers at the broad strategic scale to indicate the likely risks associated with existing activities and future developments (see Table 10.5). However, as the maps generally have a non-statutory role there are concerns over their effectiveness. Obviously all maps avoid a degree of compromise between the representation of natural complexity and ease of interpretation, and the scales used in the maps published dictate that the spatial resolution is often coarse. Thus whilst a useful strategic tool, individual sites need to be considered by more detailed assessments appropriate to the issues of local importance. Moreover, vulnerability maps need to be used in a balanced way to protect against the most significant risks to groundwater integrity, as total protection of all major aquifers would unrealistically restrict many activities in a manner that would probably be unacceptable in economic terms and infeasible.

Table 10.3 Summarised vulnerability classification for UK groundwater. After NRA (1992), BGS (1994), ADRIS (1995)

Vulnerability	Geological class	Soil class
1	Highly permeable: Chalk and Upper Greensand, Jurassic limestones, sandstones, Magnesian limestones, Carboniferous limestones, Permian and Triassic sandstones, some Lower Carboniferous sandstones, most Devonian sandstones.	High leaching potential: Soils which rapidly transmit liquid discharges due to shallow depths or extensive macroporosity. Deep coarse-textured soils with high permeability and low attenuation potential. Free passage of non-absorbed contaminants. Includes rendzinas and shallow brown earths.
2		Intermediate leaching potential: Soils where moderate attenuation is anticipated but non-adsorbed contaminants and liquid discharges can penetrate into groundwater. Includes most brown earths, brown podzolic soils and drained peats.
3		Low leaching potential: Soils where contaminants are unlikely to reach groundwater due to lateral flow (though this may cause recharge elsewhere) or a high degree of attenuation. Generally clay soils such as surface water gleys.
4	Moderately permeable: Coal measures, Millstone Grit, Old Red Sandstone, Jurassic sediments, most Carboniferous sediments, river gravels, Crag deposits, Tertiary sands and gravels.	
5	Weakly permeable: Many igneous and metamorphic rocks. All pre-Quaternary mudstones.	

SOURCE AREA PROTECTION ZONES

At the more detailed local scale, protection of individual groundwater sources becomes a priority where more stringent measures can be taken to avoid pollution. As with vulnerability mapping, this involves land surface zoning by delineating concentric areas around a particular borehole or spring. Consequently, activities in these zones can be controlled so that inappropriate development does not take place. Moreover, activities that have been sited in inappropriate places historically can be examined and relocated if necessary.

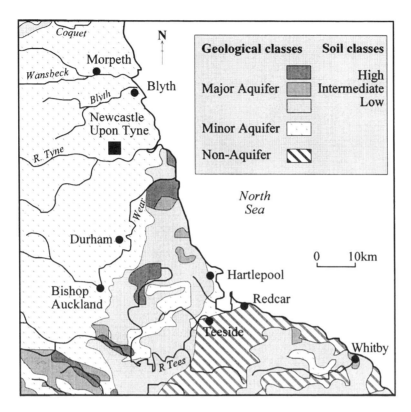

Figure 10.4   *Example of a groundwater vulnerability map for northeast England. After NRA (1992)*

Table 10.4 summarises the main features of the land surface zoning scheme that has been developed for the UK (BGS, 1991; NRA, 1992, ADRIS, 1994). The objective of this scheme is to minimise the risk of contaminating abstracted water supplies using a knowledge of the average flow velocities of groundwater which determine the travel time of pollutants to water supplies. As with vulnerability mapping, this involves a generic approach and utilises notions of simple advective transport of pollutants rather than incorporating dispersion or chemical processes which might influence the rates of transport of individual pollutants. In the immediate vicinity of the well-head, a courtyard protection zone between 10 and 100 m in diameter is recommended where low intensity land management in an area preferably owned by the abstractor is the objective. Around that, an inner source protection zone, approximating the 50-day isochron, is recommended. The 50-day isochron is identified to correspond to a time period for which pathogenic bacteria are likely to remain active (BGS, 1991). In reality, it is known that bacteria can survive for a year or more in freshwater environments; however, in reported cases of groundwater contamination the travel time is usually less than 20 days. Thus 50 days represents a conservative precautionary approach. The size of the area bounded by the 50-day isochron obviously depends upon aquifer properties. Thus, Figure 10.5 demonstrates conceptually the differing situations of relatively

Table 10.4   *Source protection zones identified for UK groundwater abstractions*

Zone	Purpose	Criteria
Courtyard	To prevent direct contamination due to pollution ingress in and around the well-head.	Arbitrary 30 m radius around the well-head where all activities not associated with the water abstraction are prohibited. Ideally should be owned and controlled by the abstractor.
Inner protection zone	To minimise risk of pollution from pathogenic bacteria and viruses.	50-day isochron, dimensions should be a minimum of 100 m to a maximum of 2 km depending upon aquifer properties.
Outer protection zone	To minimise risk of pollution by degradable toxic chemicals and to permit remedial action in the case of a major pollution incident.	400-day isochron.
Capture zone	To control persistent diffuse pollutants, particularly nitrate from agricultural activities.	Recharge area of source based on abstraction design (and licensed) yield, recharge rate and groundwater flow patterns.

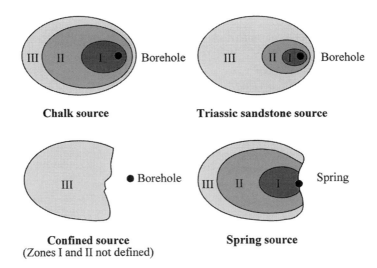

Figure 10.5   *Conceptual diagram of source protection zones in different hydrogeological settings. After NRA (1992)*

large zones for highly permeable aquifers such as chalk where the travel times are rapid, compared to more moderately permeable aquifers such as Triassic sandstone where the inner zones are smaller, reflecting the slow velocities. As a general rule of thumb, the diameter of the inner protection zone should be not less than 100 m and no more than 2 km and occupy at least 25% of the capture zone (BGS, 1991).

Beyond the inner protection zone, the 400-day isochron bounds the outer protection zone. A travel time of 400 days is a general guideline to allow a reasonable period for the degradation of toxic substances, though it is recognised that different pollutants degrade at different rates (Apello and Postma, 1993). However, for persistent pollutants, the 400-day isochron at least allows a year or so for remedial action should serious pollution occur. The total catchment of a particular water source needs to be identified beyond the outer protection zone so that as far as possible all the pollution risks within the groundwater catchment can be known.

Figure 10.5 also demonstrates the differing situations of confined aquifers and spring sources that can arise and the difficulties of zoning around confined sources in particular. This is because the 50- and 400-day isochrons may lie within the confined layer and be difficult to delineate, though clearly confining beds will limit pollution migration.

As with vulnerability maps, the source area protection concept depends upon the competent authority having the power to restrict or prevent certain activities (Table 10.5). For example, anything that requires a consent to discharge can be regulated according to its position within a source's protection zone (Table 10.5). Where powers to control certain activities are limited, the responsibility may be to persuade and educate the authority (e.g. planners) or individuals (e.g. agriculture) who are directly responsible for land management activities. In some cases restrictions may require payment of compensation.

A major issue in identifying source protection zones is the approach used to define them. A range of techniques have been applied in differing situations in different countries. In order of increasing accuracy and expense, these include: the use of fixed-radius protection zones; the use of generalised elliptical shapes around a borehole, the dimensions of which are based on aquifer properties; simple analytical groundwater models; and complex, distributed numerical flow models (IHP, 1991). Cost is a major constraint and developing a numerical model of a site might require over 200 man hours of a senior hydrogeologist's time and may cost a hundred times more than using fixed-radius areas which could be delineated by non-specialist staff (USEPA, 1987). Whilst the increased time and expense would improve the accuracy of the source area identification, it would not involve removing all of the uncertainty that is inherent in hydrological modelling. The EA in England and Wales has developed an approach base on simple steady-state modelling using readily available groundwater flow models. A prioritised programme was developed to delineate protection zones for the 1500 or so potable water abstractions in England and Wales (NRA, 1992). A final important aspect of source protection zones is developing an adequate monitoring network around the well-head in order to identify potential pollution problems at an early stage so that remedial action can be taken promptly.

Table 10.5 *Examples of general land use restrictions and guidelines in source protection zones and resource protection. After NRA (1992), ADRIS (1995)*

(a) Source protection

Activity	Inner zone	Outer zone	Catchment zone
Application of strong organic sludge	Not acceptable	Not acceptable	Only acceptable after site evaluation
Septic tank discharge to underground strata (large discharge)	Not acceptable	Presumption against	Only acceptable after site evaluation
Septic tank discharge to underground strata (small discharge)	Not acceptable	Acceptable (subject to site evaluation)	Acceptable (subject to site evaluation)
Industrial effluent discharge to underground strata	Not acceptable	Presumption against	Acceptable (subject to investigation)

(b) Resource protection

Activity	Major aquifer	Minor aquifer	Non-aquifer
Application of strong organic sludge	Presumption against	Presumption against	Acceptable (subject to site investigation)
Septic tank discharge to underground strata (large discharge)	Acceptable (subject to site investigation)	Acceptable (subject to site investigation)	Acceptable (subject to site investigation)
Septic tank discharge to underground strata (small discharge)	Acceptable (subject to site investigation)	Acceptable (subject to site investigation)	Acceptable (subject to site investigation)
Industrial effluent discharge to underground strata	Presumption against	Acceptable (subject to site investigation)	Acceptable (subject to site investigation)

GROUNDWATER PROTECTION AT THE SPEY ABSTRACTION SCHEME

Background

To illustrate the technical difficulties in developing groundwater protection strategies from a hydrological perspective, we present here a description of work carried out to develop source area protection zones at the Spey abstraction scheme in Scotland. The practical and economic constraints which then dictate the implementation of a protection strategy highlight some of the difficulties in using the output from scientific models.

Study Site

The largest public groundwater abstraction in Scotland utilises the alluvial gravels of the River Spey floodplain near Fochabers (Figure 10.6a and b). A well-field of 36 boreholes runs adjacent to the west bank of the River Spey over a distance of 3 km (Figure 10.6b). The well-field has a design yield of 27 Ml/d with 32 wells in production at any instance in time. The boreholes are approximately 50–100 m from the river and are around 100 m apart. These are drilled into the alluvial aquifer of the floodplain which is 10–12 m deep and overlies the Old Red Sandstone beneath (Figure 10.7). The design yield of the scheme is produced by groundwater abstraction from the alluvial aquifer and induced stream leakage from the River Spey (Watt *et al.*, 1986). The Spey is one of the largest rivers in Scotland, draining a catchment of some 3000 km^2, with a mean annual flow of 60 m^3/s. The river partially penetrates the upper layer of the alluvial aquifer; the lack of fine sediment on Scottish gravel-bed rivers allows unimpeded river–aquifer interactions (Younger *et al.*, 1993).

Conceptually the aquifer may be viewed as two layers separated by a 0.5 m thick discontinuous silt–clay layer (Figure 10.7). The upper layer is an unconfined aquifer and the lower layer can be considered as being semi-confined. Pump tests have shown that the saturated hydraulic conductivity of the alluvial aquifer varies between 20 and 200 m/d, with an average value of 80 m/d. The storage coefficients of the alluvium range from 9×10^{-5} in the semi-confined alluvium to 0.1 in the unconfined alluvium (Chen and Soulsby, 1996).

The groundwater that is abstracted from the scheme is drawn from the saturated zone within the floodplain which is characterised by productive agricultural soils. Thus, cereal cropping, some livestock production and market garden cropping are all practised on the floodplain. Consequently, the primary risks to groundwater are those stemming from agricultural activities. As the water table is only some 1–2 m below the soil surface the groundwater is potentially vulnerable to pollution. In addition, it is clear that any surface water pollution in the River Spey can potentially affect the quality of abstractions, though the Spey is a river of high quality (Chen and Soulsby, 1996). In order to identify source protection zones around the well-field, a numerical modelling study was carried out to identify 50-day and 400-day isochrons which could be used to delineate the inner and outer source protection zones (Sophocleous and Perkins, 1993).

Modelling Study

The readily available MODFLOW package, produced by the US Geological Survey, was utilised in this investigation to produce a steady-state, quasi-three-dimensional numerical model of the Spey abstraction scheme (MacDonald and Harbaugh, 1988). Full details of the modelling procedures are given by Chen *et al.* (1997a) and only a summary follows. Approximately 6 km^2 of the Spey floodplain surrounding the scheme was represented in a model grid with approporiate boundary conditions (Figure 10.8). No flow boundaries were set at the end and base of the alluvium, constant head boundaries were set at the northern and southern ends of the study area and a recharge boundary was specified at the surface of each cell. The river nodes were assigned fixed heads consistent with the requirements of a steady-state model.

The model was set up initially with measured values of key parameters, which were optimised in a calibration process which sought to produce a steady-state model of

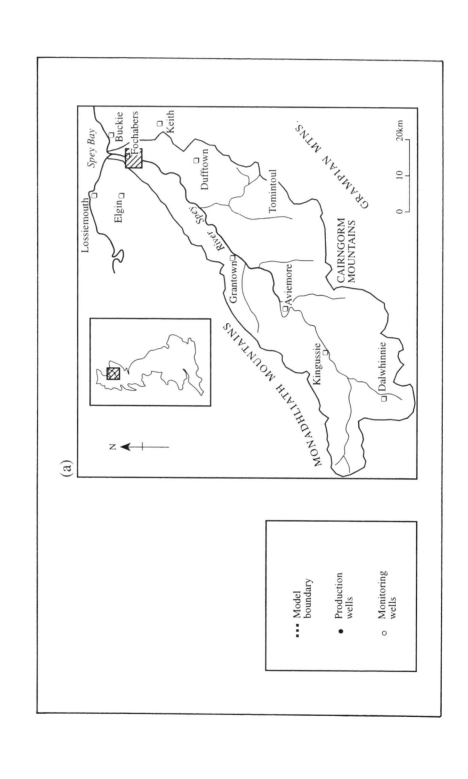

(a)

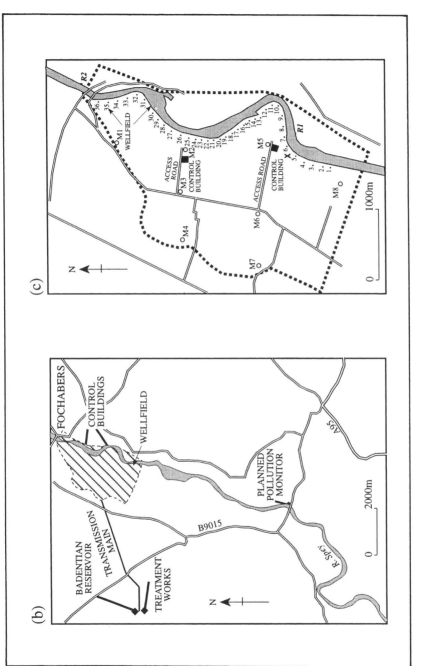

Figure 10.6 Location of (a) the Spey catchment, (b) the Spey abstraction scheme and (c) the well-field characteristics

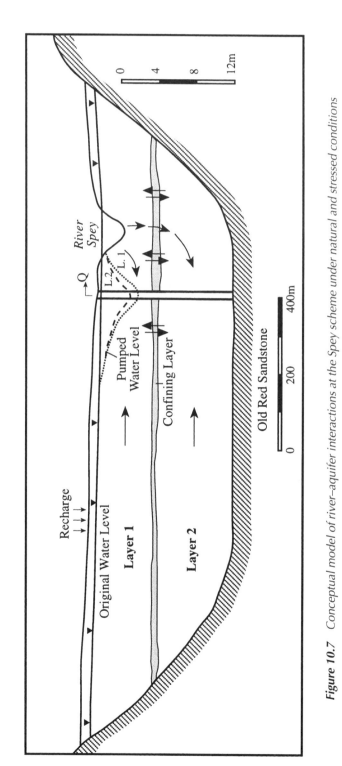

Figure 10.7 Conceptual model of river–aquifer interactions at the Spey scheme under natural and stressed conditions

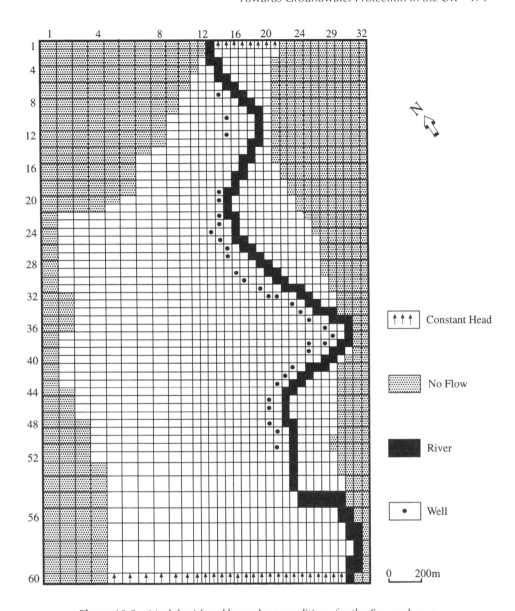

Figure 10.8   *Model grid and boundary conditions for the Spey scheme*

groundwater behaviour under unstressed conditions (details are given in Chen *et al.*, 1997a). Observations made as part of the feasibility studies for the Spey scheme showed that regional groundwater flow was from the southwest to the northeast along a hydraulic gradient of around 0.003. A good agreement between measured hydraulic heads and simulated heads was achieved (Figure 10.9). Under natural conditions the model accurately predicts the behaviour of the river–aquifer system showing groundwater flow into the river under the prevailing hydraulic gradient (Table 10.6).

When the model was used to simulate behaviour under stressed conditions, with the scheme operating at design capacity of 27 Ml/d, the river–aquifer interactions are reversed with river water leaking into the surrounding alluvium under the hydraulic gradient imposed by drawdown around the borehole (Figure 10.10). The drawdown was predicted to exceed 5 m at some boreholes, a prediction that was corroborated by observations when the scheme operated at full pumping capacity during a trial run in August 1995.

Delineating Source Protection Zones

As part of the MODFLOW suite of programmes, a reverse particle tracking package MODPATH can be used to identify time-related travel zones (Barlow, 1994). This computes flow paths under design operation to calculate linear velocity fields from the simulated head distribution. The particle tracking over 50-day and 400-day periods was achieved by introducing particles around each borehole and reverse tracking over these pre-specified time periods (Blair *et al.*, 1990).

The capture zones identified for source protection reveal complex sizes and shapes largely determined by the hydrogeology of the river–aquifer system and the direction of regional groundwater flow (Figure 10.10b). Owing to the influence of the river in sustaining groundwater abstractions, the 50-day isochron is relatively restricted around each well-head. Owing to the high water-transmitting properties of the alluvium and the complex hydrogeological system that the model represents, the capture zone has irregular shapes. These consist of discrete regions relating to the distance between neighbouring wells and their respective yields. The extent of the zone is generally restricted to within 200 m of the river, but the cumulative size is approximately 0.6 km² or 10% of the modelled area.

The 400-day isochron also exhibits a highly irregular geometry around the well-field. Its extent covers around 0.75 km² extending on both sides of the river. As the river contributes to the abstraction, the total source catchment includes all of the Spey catchment upstream. However, all of the 6 km² floodplain area is included in the catchment, but a 5-year isochron is drawn on Figure 10.10b to show the relatively slow travel times in the outer region of the floodplain for source protection purposes.

Implementing Groundwater Protection Strategies at the Spey Scheme

Given the agricultural land use on the Spey floodplain in the vicinity of the scheme, identifying protection zones for the alluvial aquifer is important. However, the design

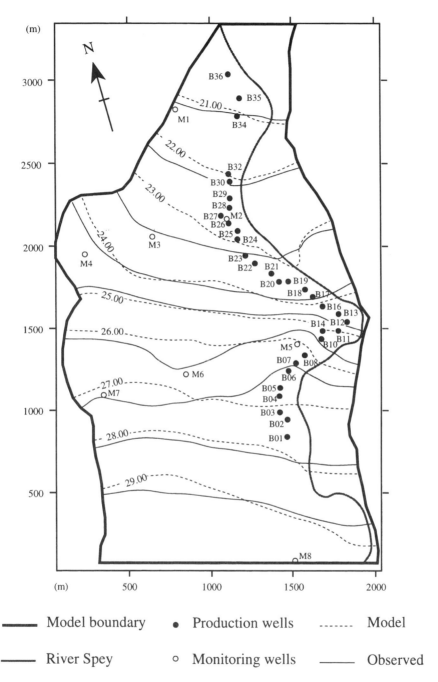

Figure 10.9 *Observed groundwater contours (m) for the Spey floodplain and predicted contours from a steady-state model*

Table 10.6 *Groundwater–surface water interactions at the Spey abstraction scheme from simulated water budgets*

Sources and discharges	Flow (m³/d)	Percentage of total	River–aquifer interaction
(a) *Pre-development*			
Sources			
Recharge	1680	20	Groundwater flow into river
Boundary inflow	6545	78	predominates
Seepage from river	198	2	
Total	8423	100	
Discharges			
Seepage to river	7914	94	
Boundary outflow	506	6	
Total	8420	100	
(b) *Design abstraction* (27 Ml/d)			
Sources			
Recharge	1680	6	
Boundary inflow	7941	27	
Seepage from river	20 094	67	
Total	29 715	100	74% of abstraction is
			derived from river
Discharges			
Seepage to river	2409	8	
Boundary outflow	367	1	
Well-field abstraction	27 000	91	
Total	29 716	100	

and development of the scheme since the early 1980s pre-dates the recent developments in aquifer protection policy that are described above. In this particular instance therefore, the abstractor, the North of Scotland Water Authority, actually took the exceptional measure of purchasing the farmland in the Spey floodplain within the modelled area to ensure full control on activities within this area. This decision to purchase was based on recognition that it was vital to preserve the integrity of a major public water supply derived from the aquifer. Thus management agreements between the Authority and the tenant who farms the land restrict the spatial extent and timing of certain agricultural practices. Thus spraying of persistent pesticides is prohibited by mutual agreement, as is slurry spreading. Ploughing is restricted to certain times of the year to minimise nitrate leaching. A monitoring network of observation wells has been installed on the floodplain to identify rapidly any changes in groundwater quality (Figure 10.6). Despite these constraints, economically successful farm management is possible. However, some of the management decisions are also based upon political considerations, in that public perceptions of activities such as slurry spraying, even if well beyond the 400-day isochron, would be perceived to be very close to the abstraction source. Thus, particularly given the increasing costs that consumers expect, and indeed are willing to pay for cleaner water supplies (Schultz and Lindsay, 1990), such perceptions are treated sensitively.

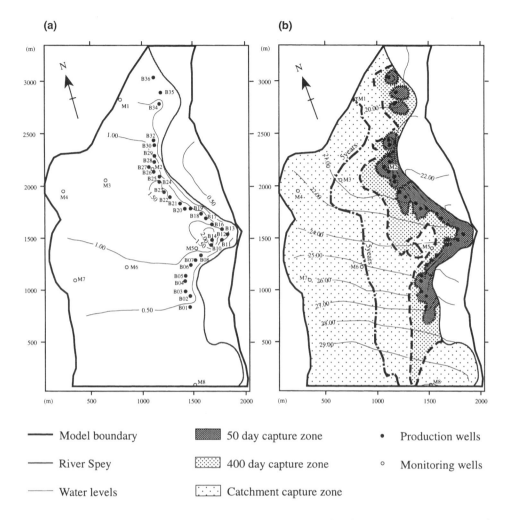

Figure 10.10 *(a) Simulated water table drawdown (m) under design pumping regime with (b) source protection zones superimposed*

This raises a general issue with regard to groundwater protection in relation to source area protection zones being meaningful in the context of land management decision making. For example, whilst knowledge of the travel-time zones is clearly advantageous from a management perspective, farm management plans are based upon field boundaries, not elliptical isochrons. Thus, interpreting and presenting travel zones in a way that is meaningful for land managers or other decision makers is important if non-specialists are to be able to interpret and implement groundwater protection strategies. Furthermore, potential errors in the modelling dictate that a precautionary approach is always advisable (Lerner and Kumar, 1991). Whilst the ownership of the Spey scheme allows the North of Scotland Water Authority to be particularly generous in this respect, land owners around many other groundwater sources are likely to be more reluctant to forego revenue from the point of view of groundwater protection. However, from the agricultural perspective agro-environment grants are now available for set-aside schemes which have considerable potential to be integrated into groundwater protection strategies by taking certain areas out of production (Potter *et al.*, 1993).

CONCLUSIONS

This example of the Spey abstraction scheme reveals a number of the difficulties in integrating scientific understanding with policy development in groundwater protection strategies. Whilst strategic resource protection is possible on the grounds of identifying levels of vulnerability associated with groundwater resources with differing hydrogeological characteristics, the degree to which this can be usefully translated into protection policies depends upon political and economic constraints that will usually require some degree of compromise. The same issues are pertinent to operational protection of individual groundwater sources. Here a range of scientific approaches ranging from simple fixed-radius protection zones to advanced contaminant transport modelling can be used to delineate source protection zones. In reality their implementation requires judgement, in addition to whatever approach is used in order to incorporate adequate safety margins at an appropriate economic cost.

The values of groundwater protection in sustainable water resource management has undoubtedly been recognised, given the recent initiatives in the UK and the rest of Europe to develop appropriate policies. However, there is still concern in the UK that the competent authorities in the form of EA and SEPA have responsibility without power, given that in many cases successful implementation depends upon persuasion and education and actual statutory powers are relatively weak compared to other countries such as The Netherlands (Jones, 1997).

Nevertheless, it is perhaps the global aspect of groundwater protection that creates the most urgent pressure for the development and implementation of groundwater protection strategies. Hiscock (1994) outlines the lack of protection strategies in many developing countries that compromise the integrity of public water supply wells. The problem is particularly acute in semi-arid areas where groundwater is the dominant source of water supply. It thus behoves the wealthier countries of the world, such as the UK, to show vision in implementing protection strategies and developing improved approaches. This would be in the long-term national interest and, in addition to providing opportunities for technology transfers, would provide a responsible management model for developing countries. As two billion people do not have access to clean drinking water supplies at the present time, preserving the integrity of existing high quality groundwater supplies must be a priority for the international community.

ACKNOWLEDGEMENTS

The authors are grateful to the following: the Bank of Scotland and the University of Aberdeen Research Committee for funding this work; staff at the North of Scotland

Water Authority who freely gave access to data on the Spey scheme for our use and generously provided advice and comments; Alison Sandison and Jenny Johnston drew the figures with speed and care. The views expressed are those of the authors.

REFERENCES

Adams, B. and Foster, S.S.D. 1992. Land surface zoning for aquifer protection. *Journal of the Institute of Water and Environmental Management*, **6**, 312–319.

ADRIS (Association of Directors and River Inspectors of Scotland). 1995. *Groundwater Protection Strategy for Scotland*. ADRIS, Scotland.

Andersen, L.J. and Gosk, J. 1987. Application of vulnerability maps. *Proceedings of the International Conference on Vulnerability of Soils and Groundwater to Pollution*, Noordwijk, The Netherlands.

Appelo, C.A.J. and Postma, D. 1993. *Geochemistry, Groundwater and Pollution*. Balkema, Rotterdam.

Arnold, G.E. and Willems, W.J. 1996. European Groundwater Studies. *European Water Pollution Control*, **6**, 11–18.

Barlow, P.M. 1994. Two- and three-dimensional pathline analysis of contributing analysis of contributing areas to public supply wells of Cape Cod, Massachusetts. *Ground Water*, **32**, 399–410.

BGS (British Geological Survey). 1991. *National Groundwater Protection Policy: Hydrogeological Criteria for Division of the Land Surface Area*. BGS Technical Report WD/91/7R.

BGS (British Geological Survey). 1994. *Groundwater Vulnerability Map of Northern Ireland*. BGS, HMSO, London.

Blair, E.S., Sheets, R.A. and Eberts, S.M. 1990. Particle tracking analysis of flow paths and travel times from hypothetical spill sites within the capture area of a wellfield. *Ground Water*, **28**, 884–892.

Bourne, R.G., Massey, S., Rolle, E. and Meighen, B. 1995. Developing comprehensive state ground-water-protection programs. *Journal of Water Resource Planning and Management*, **124**, 471–478.

Chen, M. and Soulsby, C. 1996. River–aquifer interactions in the alluvial aquifer of the river Spey: combining numerical and hydrogeochemical modelling. *Hydrogeologie*, **4**, 47–60.

Chen, M. and Soulsby, C. 1997a. Implementation of Groundwater Protection Strategies in the UK. *Proceedings of the 30th International Geological Congress*, **22**, 299–307.

Chen, M. and Soulsby, C. 1997b. Risk assessment for a proposed groundwater abstraction scheme in Strathmore, North East Scotland. *Journal of the Chartered Institution of Water and Environmental Management*, **11**, 47–55.

Chen, M., Soulsby, C. and Willetts, B. 1997a. Modelling river-aquifer interactions at the Spey abstraction scheme: implications for aquifer protection. *Quarterly Journal of Engineering Geology*, **30**, 123–136.

Chen, M., Soulsby, C. and Younger, P.L. 1997b. Predicting the consequence of minewater pollution following the Polkemmet Mine, central Scotland. *British Hydrological Society 6th National Hydrology Symposium*, Salford. BHS, Wallingford, 4.31–4.40.

D'Arras, D. and Suzanne, P. 1993. Protecting water resources: legal and operational aspects. *Journal of the Institution of Water and Environmental Management*, **7**, 344–349.

Domenico, P.A. and Schwartz, F.W. 1990. *Physical and Chemical Hydrogeology*. Wiley, Chichester.

Falkenmark, M. 1990. Environmental management: the role of the hydrologist. *Nature and Resources*, **3**, 14–23.

Foster, S.S.D. and Young, C.P. 1981. Effects of agricultural land on groundwater quality with special reference to nitrate. In *A Survey of British Hydrogeology*. Royal Society, London, 47–59.

Foster, S.S.D. Bridge, L.R., Geake, A.K., Lawrence, A.R. and Parker, J.M. 1986. *The Groundwater Nitrate Problem*. Hydrogeological Report 86/2, BGS, Wallingford.

Henton, M.P. and Young, P.J. 1993. Contaminated land and aquifer protection. *Journal of the Institution of Water and Environmental Management*, **7**, 539–547.

Hiscock, K. 1994. Groundwater pollution and protection. In O'Riordan, T. (Ed.), *Environmental Science for Environmental Management*. Longman, London, 246–262.

Howarth, W. 1990. *The Law of the National Rivers Authority*. Centre for Law in Rural Areas, Aberystwyth, Wales.

Hughes, J. and Heathwaite, A.L. 1995. *Hydrology and Hydrochemistry of British Wetlands*. Wiley, Chichester.

Hyde, G., Pedersen, K.E. and Anderson, E. 1994. *Aquifer-wide groundwater restoration and protection; from concept to practice, an example from Denmark*. IAHS Publication Number 220, 471–478.

IHP (International Hydrological Programme). 1991. *Integrated Land Use Planning and Groundwater Protection in Rural Areas*. IHP Technical Documents in Hydrology, Paris.

Jones, J.A.A. 1997. *Global Hydrology: Processes, Resources and Environmental Management*. Longman, London.

Kovar, K. and Krasney, J. 1995. *Groundwater quality: remediation and protection*. International Association of Hydrological Sciences. Publication No. 225.

Lerner, D.N. and Kumar, P.B. 1991. Defining the catchment of a borehole in an unconsolidated valley aquifer with limited data. *Quarterly Journal of Engineering Geology*, **24**, 323–331.

Lerner, D.N. *et al.* 1993. Post-script: summary of the Coventry groundwater investigation and implications for the future. *Journal of Hydrology*, **149**, 257–272.

Lewis, W.J., Foster, S.D.D. and Draser, B.S. 1982. *The risk of groundwater pollution by on-site sanitation in developing countries*. WHO-IRCWD Report 01-82.

MacDonald, M.G. and Harbaugh, A.W. 1988. A modular three-dimensional finite-difference groundwater model. *Techniques of Water Resource Investigations*, Book 6. US Geological Survey, Chapter A1.

Mackay, R. 1991. Groundwater quality management. In Biswas, A.S. (Ed.), *Environmentally Sound Water Management*. Oxford University Press, Oxford.

MAFF (Ministry for Agriculture, Fisheries and Food). 1991. *Code of Good Practice for the Protection of Water*. MAFF, Wolverhampton.

NRA (National Rivers Authority). 1992. *Policy and Practice for the Protection of Groundwater*. NRA, Bristol.

NRA (National Rivers Authority). 1993. *Water Resources Strategy*. NRA, Bristol.

Neal, C., Robson, A.J., Shand, P., Edmunds, W.M., Dixon, A.J., Buckley, D.K., Hill, S., Harrow, M., Neal, M. and Reynolds, B. 1997. The occurrence of groundwater in the Lower Palaeozoic rocks of upland central Wales. *Hydrology and Earth Systems Science*, **1**, 3–18.

O'Riordan, T. 1994. Environmental science on the move. In O'Riordan, T. (Ed.), *Environmental Science for Environmental Management*. Longman, London, 1–11.

Potter, C., Cook, H. and Norman, C. 1993. The targeting of rural environmental policies: an assessment of agri-environmental schemes in the UK. *Journal of Environmental Planning and Management*, **36**, 199–216.

Robins, N.S. 1988. Hydrogeological provinces and groundwater flow systems in Scotland. *Scottish Journal of Geology*, **24**, 51–60.

Robins, N.S. 1990. *Hydrogeology of Scotland*. HMSO, London.

Robinson, H. and Gronow, J. 1992. Groundwater protection in the UK: assessment of the landfill leachate source term. *Journal of the Institution of Water and Environmental Management*, **6**, 229–236.

Schultz, S.D. and Lindsay, B.E. 1990. The willingness to pay for groundwater protection. *Water Resources Research*, **26**, 1869–1875.

Sheils, A.K. 1993. Hydrogeology and European legislation. *Quarterly Journal of Engineering Geology*, **26**, 227–231.

Skinner, A.C. 1991. Groundwater – legal controls and organizational aspects. In Downing, R.A.

and Wikinson, W.B. (Eds), *Applied Groundwater Hydrology*. Oxford University Press, Oxford, 8–15.

Sophocleous, M. and Perkins, S.P. 1993. Calibrated models as management tools for stream–aquifer systems: the case of central Kansas, USA. *Journal of Hydrology*, **152**, 31–56.

Soulsby, C. 1996. Hydrochemical processes. In Wilby, R. (Ed.), *Contemporary Hydrology*. Wiley, Chichester, 59–106.

Soulsby, C., Chen, M., Ferrier, R.C., Helliwell, R.C., Jenkins, A. and Harriman, R. 1997. Hydrogeochemistry of shallow groundwater in an upland Scottish catchment. *Hydrological Processes*, **12**, 1111–1127.

Soulsby, C., Moir, H., Chen, M. and Gibbins, C. 1998. Impact of groundwater development on Atlantic salmon spawning habitat in a Scottish river. In Wheater, H. and Kirby, C. (eds), *Hydrology in a Changing Environment Volume I*. John Wiley & Sons, Chichester, 269–280.

Todd, D.K., Tinlinn, R.M., Scmidt, K.D. and Everett, L.G. 1976. *Monitoring Groundwater Quality: Monitoring Methodology*. USEPA Report No. EPA/600/4-76-026, Nevada.

USEPA (US Environmental Protection Agency). 1987. *Guidelines for deliniation of wellhead protection areas*. Office of Groundwater Protection, EPA, Washington, USA.

Watt, G.D., Mellanby, J.F., van Wonderen, J.J. and Burley, M.J. 1986. Groundwater investigations in the lower Spey valley near Fochabers. *Journal of the Institution of Water and Environmental Management*, **1**(1), 89–103.

Younger, P.L., Mackay, R. and Connorton, B.J. 1993. Streambed sediments as a barrier to groundater pollution: insights from fieldwork and modelling in the river Thames Basin. *Journal of the Institution of Water and Environmental Management*, **7**, 577–585.

11

Organisational Issues and Policy Directions for Urban Pollution Management

J. Bryan Ellis and Bob Crabtree

INTRODUCTION

In common with many states, there are many environmental problems in England and Wales associated with the collection, treatment and disposal of rainfall runoff-generated stormwater. These have arisen from historical inadequacies in science, technology, management and funding. This chapter considers the advances that have occurred in recent years to reinforce and enhance regulatory, financial and technological capabilities to allow progress to be made towards meeting the current requirements of legislation and public expectation.

The last decade has seen a progressive development in technical know-how, cost-effective planning and evaluation tools for use by the UK water industry for tackling the problems associated with urban wet weather discharges. The solutions to both potential and existing problems can now be reasonably addressed from the standpoint of a good baseline understanding of sewer and river system performance. This is allowing the restricted available investment to be applied in a customised fashion to avoid excessive and unnecessary expenditure through the technical framework provided by the *Urban Pollution Management Manual* (FWR, 1994). This procedure has placed the water service companies in a position to confidently and appropriately address a backlog of improvements to water reclamation works and combined sewer overflows (CSOs). However, as yet the full extent of problems associated with untreated impermeable

Water Quality: Processes and Policy. Edited by Stephen T. Trudgill, Des E. Walling and Bruce W. Webb.
© 1999 John Wiley & Sons Ltd.

surface stormwater discharges have still to be evaluated. The difficulties and frustrations of effecting changes in urban runoff management are clearly identified by reference to the recent revision of the *Sewers for Adoption* manual (Water Research Centre, 1995). This fourth edition makes no mention at all of surface storage and infiltration systems and indeed the design criteria for CSOs has been withdrawn, thus moving more strongly towards the policy of continued discharge of untreated stormwater to receiving waters despite clear evidence of adverse impacts (House *et al.*, 1993). This has meant that developers have retained a traditional conservatism to urban drainage design by moving away from surface storage and disposal to below-ground tank storage systems.

At the same time, the regulatory authorities have espoused strategies for the sustainable management of water resources and integrated catchment planning which are consistent with the UK government position on sustainable development as stated in its 1994 policy document (Department of the Environment, 1994) and in parallel with this, local authority strategies consistent with Local Agenda 21 (LGMB, 1993). Key catchment planning issues contained within Regional Planning Guidance also place emphasis on sustainability principles for surface water quality (NRA, 1995).

This chapter contrasts the approaches to urban surface stormwater management and urban wastewater management in England and Wales. In the former case policy and practice lag behind the scientific understanding of a source control approach to pollution management. Similarly, discharge of stormwater is the responsibility of several organisations. Urban wastewater management policy is more soundly based around organisational and financial frameworks with recognised, science-based approaches to pollution control. However, future policy changes to meet increased demands for environmental improvements may result in policies which cannot be applied without further advances in the understanding of wastewater system performance, associated environmental impacts and the evaluation of potential improvement schemes.

CONTROL OF SURFACE RUNOFF POLLUTION

In terms of urban surface water management, the diversion, attenuation and disposal of impermeable stormwater discharges at source is seen by many (SEPA, 1996) as a key concept supporting sustainability as it is focused on pre-emptive prevention of downstream water flow and quality problems for both in-sewer and within receiving water bodies. In addition, it is argued that strategic source disposal of surface runoff may provide a valuable contribution to aquifer storage and recovery (ASR) although none of the current UK water company schemes yet involve direct stormwater recharge. The USA, The Netherlands and Australia have successfully harvested roof and street drainage through infiltration to groundwater as a basis for mains water replacement for secondary uses such as irrigation and recreation (Argue, 1994). Whilst rising urban groundwater levels have been widely recognised as a problem, there are considerable uncertainties as to whether long-term source disposal of urban stormwater is likely to cause widespread contamination of urban groundwaters.

Whilst the impacts of episodic discharges from impermeable urban surfaces upon receiving surface water systems are now well documented (Ellis and Hvitved-Jacobsen,

1996), there are relatively few field assessments available to evaluate actual impacts of stormwater infiltration to urban groundwaters. The potential for highway discharges to contaminate local aquifers has certainly been recognised, especially where roadside filter or fin drains can directly infiltrate to underlying fissured strata (Price, 1994; Luker and Montague, 1994). In the UK, it is now Environment Agency policy not to allow major highway and motorway discharges within designated Zone I (Inner Source Protection) regions and they will only be acceptable under exceptional circumstances in Outer Source Protection Zones.

Within the context of integrated catchment planning, regulatory authorities are increasingly identifying a range of source control techniques which are perceived as constituting a suite of Best Management Practices (BMPs) for the sustainable management of intermittent urban runoff (CIRIA, 1992). As most source control systems divert surface runoff to groundwater, it could be argued that they constitute a valuable source of aquifer recharge. The theoretical risks they present to groundwater pollution must therefore be set against the potential benefits to be gained for example from the possible strategic recharge of some $210 \, \text{Mm}^3$ per annum that could be generated from an average annual 500 mm of rainfall falling on some $416 \, \text{km}^2$ ($52\,000 \, \text{km}$) of highway surfaces laid on the chalk of southeast England.

Stormwater Infiltration and Quality Standards

The type and range of pollutants associated with urban stormwater is extremely variable (Ellis, 1986) but five pollutant groups are of principal concern in terms of the potential use of urban runoff for ASR.

Solids and Heavy Metals

A significant proportion of the total and toxic polluting load arising from urban surfaces is associated with the fine ($ < 63 \, \mu m$) particulate fraction of the discharge. Lead levels in urban stormwaters considerably exceed most current UK quality standards with the exception of waters intended for irrigation (Figure 11.1). In general, levels of other metals such as zinc, cadmium, copper and chromium fall within the guidelines. The suspended solids (SS) concentration range is typically two orders of magnitude and the 190 mg/l Event Mean Concentration (EMC) value for stormwater runoff is of significance to ASR because fine solids are a primary cause of clogging of both injection wells and infiltration systems. In addition, this fraction may contain over 90% of inorganic lead as well as 70% of the copper, chromium and hydrocarbons. Whilst settleable and insoluble materials may not lead to any immediate failure of water quality standards, they could be leached out of infiltration systems through release of mobile, colloidal particles in association with elevated dissolved organic carbon (DOC) levels (10-12 mg/l) into the underlying saturated zone.

Hydrocarbons and Pesticides

Hydrocarbons and pesticides (including herbicides) are List I substances under the terms of the EC Groundwater Directive and thus direct discharge of stormwater runoff

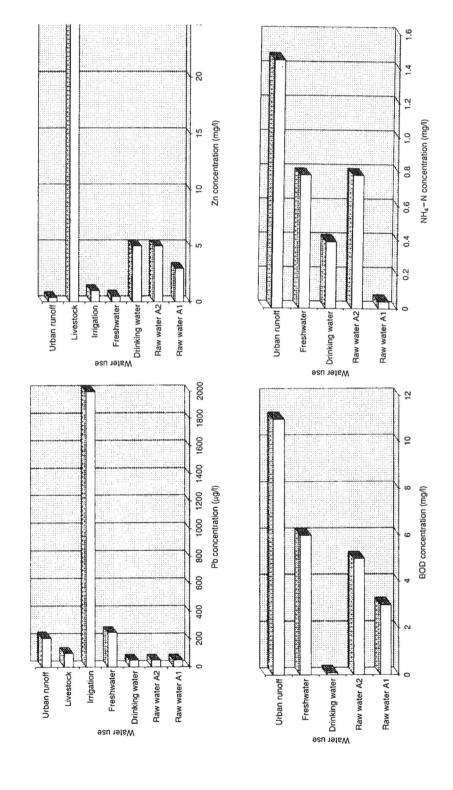

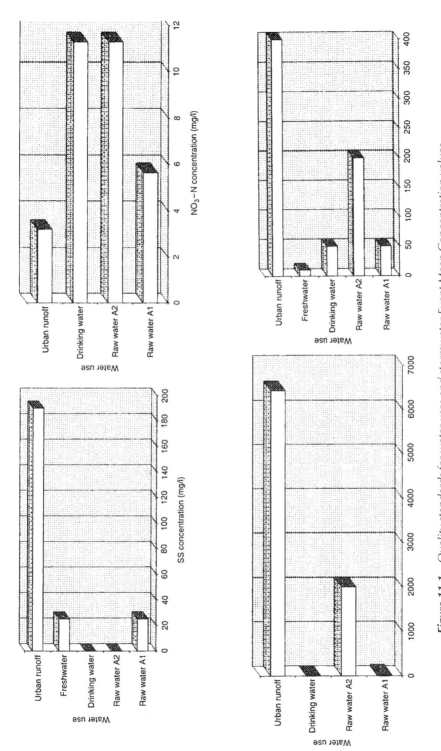

Figure 11.1 Quality standards for water uses and stormwater Event Mean Concentration values

containing these substances to groundwater is not permitted although the regulatory authorities have yet to establish Water Quality Objectives for groundwater. Until these are developed, the Environment Agency will advise on the standards which must be satisfied for individual aquifers. The Groundwater Protection Policies contained in the 1992 National Rivers Authority framework document *Policy and Practice for the Protection of Groundwater* also limit locations where indirect (via soakaways, infiltration trenches etc.) discharges are feasible, with oil interceptors required wherever either Source or Resource protection is necessary under Acceptability Matrix 3c of the policy framework. Also relevant to the disposal of impermeable urban and highway runoff are the Groundwater Policy Statements concerning diffuse pollution of groundwater (Policies G1–G4) particularly with reference to the leaching of herbicides from roadside verges and landscaped areas.

Total oil (hydrocarbon) levels in urban runoff average 10–20 mg/l (Figure 11.1) with motorway and trunk drainage averaging 25–30 mg/l and reaching as high as 100–400 mg/l during short intense storm events (Colwill *et al.*, 1984). Suburban roads have a lower range varying between 2 and 28 mg/l. The significance of such oil contamination is difficult to quantify in terms of UK legislative requirements, although Environment Agency discharge consents to surface waters of River Ecosytem Classes 1, 2 and 3 (General Quality Assessment Grades A, B and C) are in the region of 5–10 mg/l (NRA, 1993).

Bacteria

Concentrations of faecal coliform bacteria in urban stormwater are high (Figure 11.1), being significantly above the guidelines for all specified uses. Pathogenic organisms are also regularly found in impermeable stormwater runoff with *Pseudomonas* being typically at 10^2 MPN/100 ml and *Salmonella* being recovered in 60–70% of samples (Ellis, 1993). The presence of elevated DOC levels in stormwater will enhance bacterial and pathogen transport in the underlying porous media by blocking potential attachment sites.

Source Control Performance

Table 11.1 identifies a range of potential benefits and barriers to the adoption of source control methods for urban stormwater management which include organisational, technical, planning and financial as well as social and legal issues. Whilst there is a reasonable body of evidence on the performance of detention storage and wetland systems (Lawrence *et al.*, 1996), there are only limited data available on the long-term performance of other source control systems such as infiltration trenches, porous paving or grass swales.

Field Tests

Most studies that have been undertaken of infiltration system performance have shown high but erratic pollutant removal rates. This variability in performance is

Table 11.1   *Limits and benefits of urban runoff source control*

Source control type LIMITS	Retention* (no infiltration)	Detention*† (no infiltration)	Infiltration*	Recharge and/or reuse	Source control type BENEFITS
Maintenance cost and responsibility:					
Short term	1	1	1	1–2	*Reduced capital*
Long term	2	3	1	3	*expenditure:*
	A–C	*B–D*	*C–D*	*D (?)*	*Water utilities*
	C–D	*A–C*	*B–D*	*D*	*Developer*
Health and safety	1	0	0	2	
	D	*B–D*	*C*	*B–C*	*Reduced flood risk*
Capital cost	2	1	0	2	
	A	*B*	*D*	*A–B*	*Replenish baseflows*
Long-term performance	2	2	2–3	3	
	C	*C*	*B–D*	*D*	*Reduce misconnections*
Implications of system failure (connections to foul sewers)	3	2	2	3	
	B–C	*C–D*	*C–D*	*B*	*Improve river quality*
Multi-agency agreement	2	2	2	3	
	B	*B–D*	*A–C*	*A*	*Enhance wildlife habitats*
Lack of technical info/support, e.g. design criteria	3	1	2	3	
	A	*A*	*B–C*	*D*	*Reduce potable water demand*
Groundwater contamination	2	1	3	3	
	B–C	*B–D*	*B*	*A*	*Enhance development amenity value*
Legal, financial and insurance	2	1	3	3	
	A	*B*	*D*	*A*	*Enhance soil moisture regime*
Public acceptability	1	1	1–2	2	
	A–B	*C–D*	*C–D*	*A–B*	*Reduce erosion and sedimentation*
	A	*A–B*	*A–D*	*A*	*Reduce land take*

LIMITS: **0 Little or no problem; 1, some problems; 2, medium-scale problems; 3, major problems.**
BENEFITS: A, no change compared with traditional design; B, better; C, even better; D, best
* Examples. Retention: permeable surfaces (roofs, roads, car parks etc.). Detention: grass swales, wetlands. Infiltration: vertical/plane infiltration trenches, filter/fin drains, soakaways, etc.
† Storage basins/ponds not included

confirmed from a field study of a submerged aerobic biological filtration device to treat outflows from a stormwater detention basin receiving stormflows from a 440 hectare suburban catchment (Table 11.2). The filter medium used in the test was an inert expanded schist (nominal diameter 3–6 mm) which is similar to that used for *biocarbone* treatment for water and industrial wastewaters; the test was conducted under a continuous flow regime without spiking.

Net losses of nutrients occur with extended retention times and backwashed concentrations of SS and Zn were as high as 265 mg/l and 43 μg/l, respectively, suggesting that sloughed biomass and associated particulate clogging increase headloss and reduce efficiency quite rapidly. The erratic metal concentration depth profiles noted in the field tests are probably the result of variable precipitation with sulphide and adsorption onto Fe and Mn (hydr)oxides under changing redox conditions. Monthly backwashing would be required to maintain an optimum performance and, based on a 50% removal target for SS, total organic carbon (TOC), PO_4 and Pb for a 2-year, 1-hour duration design storm generating 7500 m^3 of runoff, the filter surface area would need to be about 1200 m^2. This would pose considerable space, cost and maintenance difficulties in urban areas and alongside highway verges.

Failure Rates and Design

There are general high levels of failure reported for infiltration-based techniques, with 5-year failure rates for trenches and porous pavements being 50 and 75%, respectively, for systems in the central eastern USA (Schueler *et al.*, 1992). There has likewise been a high failure rate of monolithic porous asphalt and concrete reported from the eastern USA and other countries such as Australia have abandoned this "best practice" altogether. Whilst many failures may be related to inappropriate sizing relative to catchment area or to lack of maintenance, the majority are the result of lack of solids pre-treatment and of groundwater mounding. Filter basins of, sand and other media (peat, compost, geotextiles), which have gained substantial popularity in the USA as stormwater BMPs, are also subject to high and early failure rates. Urbonas *et al.* (1996) have reported cumulative total suspended solids (TSS) removal rates falling by 70% within one year of installation with flow-through (hydraulic conductivity) rates being throttled from an initial 1 m/h to less than 0.02 m/h causing frequent and severe flow bypassing. Infiltration and filter devices must be properly sized for the expected maintenance cycle that matches both the average annual runoff volume and the

Table 11.2   *Field test filter performance*

Run number	Hydraulic retention time (h)	Surface loading rate (m/h)	Flow rate (l/min)	Removal efficiency (%)					
				SS	TOC	PO_4	NH_4	Pb	Zn
1	1.0	0.45	6.4	94	10	21	59	—	—
2	1.5	0.30	4.3	92	13	−7	97	27	66
3	2.5	0.18	2.6	89	30	−100	−88	nd	nd

nd, not detected

average annual TSS EMC in the runoff. If the control device cannot be made large enough to pass through the design event without backing-up water when it is partially clogged, sufficient stormwater detention volume or equivalent attenuation must be provided upstream to provide solids pre-treatment and to balance a clogged flow-through rate of about 10–12 mm/h. Good engineering design practice can mitigate some of the problems associated with conventional infiltration and porous paving systems but they cannot provide a fail-safe guarantee of long-term groundwater protection and neither are they able to predict a target pollutant removal rate.

For discharges to ground, oil interceptors and infiltration devices will have a negligible effect on the concentration of trace organics present in a dissolved or colloidal form. Whilst polyaromatic hydrocarbons (PAHs) are generally insoluble and solid-associated, their cosolvent properties can enhance their solubility. Monocyclic aromatic hydrocarbons such as benzene and toluene are quite mobile as are the low molecular weight phenols and in nutrient deficient aquifers, their degradation is likely to be slow. Thus unless there is specific local knowledge to indicate that degradation or attenuation will be assured, it should be assumed that none will occur.

Herbicides have become an integral part of urban management strategies for the control of vegetation by local, county, highway and airport authorities as well as rail track operators (Ellis *et al.*, 1997). The non-agricultural applications of herbicides represent 2–3% (550 tonnes) of the total amount of active ingredient applied each year in the UK. Figure 11.2 shows the exceedances of the 0.1 μg/l UK drinking water standard resulting from the use of non-agricultural herbicides in 1993 and 1994 expressed as a percentage of all pesticide exceedances by water utility regions. Herbicide losses from hard surfaces via infiltration trenches/drains now account for more

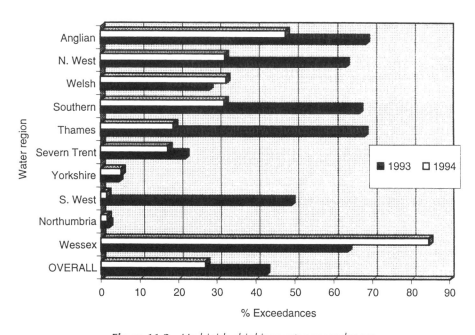

Figure 11.2 *Herbicide drinking water exceedances*

than 30% of pesticide exceedances in five regions and more than 15% in seven of the ten regions.

The current UK regulatory approach to determining the impact on water resources of such applications in urban areas is to assume that 100% of the applied product will have the potential to be washed off and thus lead to contaminated drainage water being infiltrated down to groundwater. Whilst the use of the triazine group (atrazine and simazine) was revoked in 1993, there has been a switch to other compounds, notably to diuron, chlorotoluron and isoproturon. In 1994, the latter was responsible for over 50% of all UK exceedances of the drinking water standard compared to only 7% and 2% for atrazine and simazine, respectively. The transport mechanisms and underlying governing factors that determine the movement and fate of herbicides from hard surfaces to groundwater are still very poorly understood. Until this situation changes allowing the formulation of more robust impact risk assessments, the continued use of herbicides in urban areas will remain a significant issue for groundwater protection.

Grasscrete paving and sand filters have been shown to remove between 40 and 60% of bacteria from stormwater (Raimbault and Balades, 1987) but secondary biological treatment through grass swales and/or wetland systems would be needed to provide a full groundwater protection. US Environmental Protection Agency regulations set aside the natural degradation of pathogens in aquifers and require full tertiary treatment and disinfection prior to recharge where stormwater is to be reclaimed for use as a potential drinking water supply (National Research Council, 1994). In Europe, it is common practice to allow a minimum 50-day residence time of recharge water in aquifers to enable die-off of pathogens but survival times and resuscitation mechanisms for enteric organisms in groundwater are still uncertain.

Groundwater Assessment

For discharges to ground, the annual loading resulting from surface runoff is likely to be of greater importance than the concentration discharged by any one event (except where spillages are concerned). A tiered approach is recommended for assessing the groundwater pollution potential of impermeable urban drainage. At the first stage, the methods published by the Construction Industry Research and Information Association (Luker and Montague, 1994) can be applied. In this method, an annual build-up rate for pollutants accumulating on the impermeable surface is assumed according to the appropriate land use and average traffic volume. All of the pollutant material is assumed to be washed off the surface, diluted by the total annual average effective rainfall (under an assumed runoff coefficient of 0.5) for the site to derive a calculated runoff concentration for specified pollutants. It is also further assumed that there is no removal or attenuation by the soakaway or infiltration device and that the groundwater is static, i.e. there is no flow through the aquifer.

There are a number of conservative assumptions in the approach such as:

- that pollutant build-up rates provide a reasonable surrogate for likely discharge and infiltration quality;

- that the balance between soluble and solid-associated species remains stable between initial surface mobilisation and final percolation into the saturated zone;
- that no dilution, dispersion or attenuation of pollutants occurs on passage through the drainage system, the infiltration device or the unsaturated zone.

However, given the greater vulnerability in terms of the potential and costs of remediation of groundwater to pollution, the conservative assessment is appropriate. Where the calculated concentration in the runoff is greater than the relevant environmental quality standard or specified guideline values, more detailed evaluation will be required to determine whether, and what, additional pollution control measures are needed. Table 11.3 provides a first-order screening of the removal efficiencies afforded by various treatment systems that can be considered for urban drainage intended for groundwater recharge together with estimates of capital and maintenance costs. However, the design approach should adopt a "treatment train" combination of two or more source control systems to provide sufficient safeguard against diffuse groundwater pollution from impermeable urban runoff.

URBAN POLLUTION MANAGEMENT IN ENGLAND AND WALES

In the UK, an integrated wastewater planning procedure has been developed to allow the rational investigation of pollution from urban wastewater discharges in response to rainfall events. This procedure, called the Urban Pollution Management (UPM) Procedure (FWR, 1994), enables wastewater engineers and environmental planners to identify environmentally acceptable, cost-effective solutions to urban wastewater pollution problems, and in particular to problems associated with wet weather conditions. In England and Wales, the Environment Agency's policy requires the application of the UPM Procedure to meet the requirements of the EU Urban Wastewater Treatment Directive (NRA, 1993) in planning wastewater improvement schemes to limit the risk of transient pollution from combined sewer overflows and wastewater treatment plants.

Table 11.3   *Urban stormwater drainage treatment systems*

Treatment system	Capital cost (£1000)	Maintenance cost (£/pa)	Removal efficiency (%)					
			Zn_{tot}	Cu_{diss}	Fe	Pb	SS	HCs
Kerb/gully/pipe system	180–220	1000			10–30			
Oil interceptors	10–30	300	30–50	< 10	30–40	40–60	30–80	40–80
Combined filter/French drains	160–200	—	70–80	10–30	80–90	80–90	80–90	70–90
Infiltration basin	20–50	2500	70–80	10–30	80–90	80–90	60–90	70–90
Sedimentation lagoon	60–100	2000	60–80	20–30	90+	80–90	60–90	70–90
Detention pond	15–30	350	30–40	< 10	30–50	40–60	40–70	30–60
Grass swales	15–40	350	70–90	50–70	90+	80–90	60–90	70–90
Wetland systems	15–60	2500			50–80			
Sedimentation tank	30–60	300	30–50	< 10	30–40	40–60	30–80	40–80

A sewer system has three fundamental objectives. These are:

- to convey wastewater to a suitable point for treatment and disposal;
- to drain paved areas; and
- to safeguard receiving waters from pollution.

Clearly, these objectives are not absolute but must be set as a compromise between acceptable levels of performance and cost. Traditionally, the design of sewerage systems has concentrated on the first two objectives. The requirement to protect receiving waters from pollution is an aspect of increasing attention as awareness of pollution issues has driven national and EU legislation.

Approximately 30% of UK sewer systems are combined systems which carry both foul sewage and surface runoff. Only 10% are considered to be truly separate systems. The remainder are a mixture of both types. Some 95% of the UK population of 51 million are provided with sewerage systems and 83% of the resulting wastewater receives secondary treatment. Wastewater pollution of watercourses in the UK is concentrated around the areas of older industrial and urban development, particularly in the north (Manchester, Leeds) and the Midlands (Birmingham). Considerable investment has been made in the last 20 years to reduce pollution from inadequate sewage treatment works and there has been a considerable improvement in water quality in these areas. With this general improvement, the effects of wet weather discharges and, in particular, the effects of CSOs, discharges from separate storm-water-only systems and discharges from storm sewage storage tanks at sewage treatment works, have begun to be fully appreciated.

CSOs are constructed for two reasons:

- it is uneconomic to build a pipe system capable of conveying all storm flow to treatment;
- it is uneconomic to construct treatment works capable of treating the whole of the storm flows.

Historically, in the UK, CSOs have been designed to discharge at fixed multiples of the dry weather flow (DWF) in the sewer, usually 6 × DWF. Flows in excess of 6 DWF would be spilled and the remainder carried forward to the sewage treatment works where a treatment capacity of 3 DWF is provided. The balance is stored in storm tanks for treatment later. An emergency overflow is normally provided at the storm tanks for storms of exceptionally long duration.

Problems arising from this pragmatic approach were recognised as early as the 1960s when a new approach was proposed to calculate an acceptable carry-forward flow from a CSO which took account of the balance between industrial effluent and domestic sewage (HMSO, 1970). This has been widely adopted and is still in use as a minimum sewerage performance criterion (NRA, 1993) despite its lack of consideration of receiving water impact. In the 1960s new designs of overflow were adopted to provide for retention of pollution in the sewer, such as the use of storage, stilling pond or high side weir overflows. However, increased urban development has led to overloading of existing sewer systems and a deterioration in the performance of CSOs.

Environmental Requirements for Controlling Wet Weather Discharges

There are believed to be some 25 000 CSOs in the UK. Around 30% of these (8000) have been identified by the Environment Agency as having unsatisfactory performance on the basis of the following criteria:

- significant visual or aesthetic impact due to solids, fungus in the receiving water or a history of justified public complaint;
- makes a significant contribution to a deterioration in river quality;
- makes a significant contribution to a failure to comply with the quality standards set by the EC Bathing Waters Directive Quality Standards for identified bathing waters;
- discharges in dry weather conditions;
- causes a breach of water quality standards.

It is widely recognised that it is particularly difficult to manage such discharges in an environmentally sympathetic manner and that the demand for funds is potentially very great. It is estimated that some £2 billion will be required to meet the stormwater requirements of the Urban Wastewater Treatment Directive (Clifforde *et al.*, 1995). This high level of investment must result in significant improvements to the environment and must be cost-effective.

Since privatisation of the water industry of England and Wales in 1989, the Environment Agency (and from 1989 to 1996, the National Rivers Authority) has done much to develop and implement a comprehensive, objective and logical national policy for wet weather pollution management.

The conceptual approach to stormwater management that has been adopted is one of environmental quality standards linked to use-related objectives. Emission control methods have been promulgated only where the environmental objectives approach is not feasible (e.g. aesthetic pollution control) or scientifically unjustified (e.g. Red List substances), or to offer a simpler alternative where the scale of the problem does not justify the cost or complexity of the impact assessment study necessary to implement a use-related solution (Tyson, 1991; Matthews, 1995).

Three major uses have been identified as being potentially affected by intermittent wet weather wastewater discharges.

1. River Aquatic Life, where short periods of low dissolved oxygen and/or high unionised ammonia can hinder the development of a sustainable fishery in inland waters.
2. Bathing, where frequent and persistent high concentrations of bacteria can cause non-compliance with the EU Bathing Waters directive.
3. General Amenity, where sewage solids can lower the perceived quality of the receiving water body, resulting in public complaints.

Environmental quality standards have been developed for River Aquatic Life based on tabulated intensity/duration/frequency relationships for dissolved oxygen and unionised ammonia. A simplified example of this type of standard is given in Table 11.4.

Table 11.4   *Example fundamental intermittent standards for dissolved oxygen (DO) concentration/duration thresholds not to be breached more frequently than shown. From FWR (1994) with permission*

Return period	DO concentration (mg/l)		
	1 h	6 h	24 h
1 month	4.0	5.0	5.5
3 months	3.5	4.5	5.0
1 year	3.0	4.0	4.5

The environmental standards for the protection of bathing waters are well known from the EC Directive. The acceptable duration of non-compliance due to storm discharges is computed to be 1.8% of the bathing season. An alternative emission-based approach is available in the form of a simple spill frequency criterion, expressed as not more than three spills per bathing season on average (NRA, 1993).

The bathing water criteria reflect the principle that where an emission-based alternative to an environmental quality standard is offered, the emission standard always incorporates a considerable factor of safety in terms of the environmental performance of the resulting scheme. This is to allow for the "indirect" nature of the control criteria relative to the use which is to be protected. Hence, there is always a trade-off between reduced investigative costs and increased capital cost for schemes designed in this way.

Organisation and Financing of Urban Water Quality Improvements

One of the major reasons for the privatisation of the water industry in 1989 was to free the industry from the constraints and limitations of the public purse (Tyson, 1991). The resultant utility companies are, by their very nature, monopolies within their geographical areas. Hence, in addition to environmental regulation, a strict form of economic regulation is also necessary to represent and protect the interests of the customers. This role is performed by the Office of Water Services, Ofwat.

Ofwat stipulates the amount of money above (or below) the rate of inflation that each company can raise from their customers to carry out their business over the following 10 year period. Ofwat also stipulates the performance targets for the companies over that period. Hence, in effect, both the inputs (money) and outputs (performance targets) are fixed. To arrive at these figures, Ofwat reviews a company's performance over the preceding period and considers both ongoing obligations and any new commitments that it may have for the coming period. The last 5 yearly review, which came into effect in April 1995, took into account the obligations of the companies to meet the requirements of the Urban Wastewater Treatment Directive in addition to the residual part of the Bathing Waters Directive construction programme. Countering these pressures for increased income was the demand from Ofwat for improved operational and capital efficiency from the companies.

The companies are required to submit detailed strategic business plans, called Asset

Management Plans, to Ofwat to assist in the periodic review. These Plans describe the schemes to be undertaken during the planning period in some detail. In respect of storm discharges, the most recent Plans identified the unsatisfactory CSOs to be addressed in the 10 year period commencing April 1995. It was not considered feasible to tackle all problem CSOs in this period and, in practice, the funds allocated are anticipated to allow approximately 60% of the total number of unsatisfactory CSOs to be rectified, representing those CSOs where the greatest environmental benefit will accrue from the available investment. The remaining 40% will be improved during subsequent investment programmes.

The Urban Pollution Management Manual Procedure

Having identified the environmental needs for stormwater management and the financial framework within which funds are made available to the water industry, the third essential component is a technical planning framework to allow problems to be addressed in a way that ensures environmentally effective solutions are identified, whilst avoiding excessive expenditure through unnecessary over-design. Such a framework, developed under the UPM research programme (Clifforde and Murrell, 1993), is described in a single holistic planning document, the *Urban Pollution Management Manual* (FWR, 1994).

The complete planning methodology is based on a single procedure into which modelling tools of varying type and complexity can be fitted to suit the specific local circumstances. The overall framework, illustrated in Figure 11.3, shows the key steps in moving through problem identification to data collection and model building, via an iterative solution-developing process, before the final consenting and detailed design aspects are considered.

Five key issues underlay the application of the UPM Procedure.

1. The variability of discharge loads and their associated environmental impacts is large and result from many complex processes and interactions.
2. It is necessary to take a holistic approach to solving problems in which the wastewater system and receiving water are considered as a single integrated system.
3. Limited financial resources require individually tailored solutions to pollution problems to avoid excessive and unnecessary expenditure.
4. Such solutions can only be identified from the standpoint of an adequate understanding of system performance which comes from the use of appropriate modelling tools.
5. A hierarchical system of models allows an incremental approach to problem solving and aids the identification of cost-effective, integrated solutions.

The water quality simulation models used to support the UPM Procedure are described in detail in the Manual (FWR, 1994). These include models for rainfall input generation, sewer system flows and pollutant concentrations, treatment works performance and river/marine impact assessment.

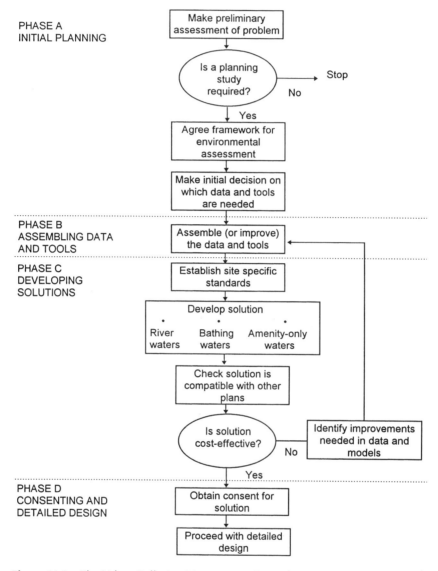

Figure 11.3   *The Urban Pollution Management Procedure. From FWR (1994) with permission*

The benefits of the UPM Procedure are:

- more cost-effective solutions which will result in overall cost savings for the water industry;
- more consistent plans since the procedure is founded on a single set of environmental criteria;
- more rapid agreement on discharge consents as common understanding will speed the process;

- familiarity and confidence in the technology which will ensure consistent and effective solutions which will meet environmental needs.

The need to demonstrate the environmental and financial benefits to be realised from the practical application of the UPM Procedure to a major catchment was considered to be an important step in promoting its uptake. Hence, a large-scale demonstration project, jointly funded by the water industry and the environmental regulators, was undertaken (Crabtree *et al.*, 1995). Derby, a large town in the Midlands of England, was selected as a suitable site as there were long-standing concerns about flooding and CSO pollution problems. Both the size (220 000 population equivalent) and complexity of the sewer system, with 23 unsatisfactory CSOs, and the critical nature of the multiple uses of the River Derwent to which the wastewater system discharges, made Derby a challenging case study for the UPM Procedure. The project was carried out by the Water Research Centre with the support of Severn Trent Water Ltd and Midlands Region of the National Rivers Authority, between the autumn of 1993 and the spring of 1995.

The Derby UPM study involved intensive data collection to support the calibration and verification of dynamic water quality simulation models for all wastewater system components. Environmental impact assessment against wet weather water quality criteria for a range of upgrading schemes indicated that schemes based on traditional planning procedures would have involved unnecessary capital expenditure and a high degree of uncertainty in meeting water quality objectives in the River Derwent (Crabtree *et al.*, 1996a).

FUTURE DEVELOPMENTS

Considerable caution is needed in advocating the widespread introduction of dispersed source control techniques for impermeable urban stormwater management where there is likelihood of long-term infiltration of such drainage waters to groundwater. Urban runoff is certainly not suitable for direct injection into aquifers to generate drinking water supplies, and pre-treatment in detention or wetland basins is recommended for infiltration to be used for secondary irrigation or livestock supplies. Particular constraints on the sustainable reuse of stormwater from source disposal arises from trace organics (especially herbicides), pathogenic bacteria and possibly from dissolved metals and hydrocarbons. Much further evidence is required to characterise, quantify and predict their rates of degradation in both infiltration systems and within the saturated zone before these source control devices can be advocated for groundwater recharge. Uncertainty regarding long-term performance and the lack of indicators for system failure are major barriers to the adoption of source control procedures as well as perceived difficulties regarding maintenance responsibilities. At present the onus for introducing alternative source control drainage lies with the developer and local authorities given that the water service companies are sceptical of their benefits and argue that there are legal difficulties in adopting controls such as infiltration systems. The recent Environment Agency inventory of aquifer pollution in England and Wales resulting from point sources such as landfill and the legal difficulties of implementing retrospective remedial measures, provides a clear warning of the dangers of similarly conceding to widespread source disposal of contaminated urban stormwater to groundwater.

Recent changes to the organisation and funding of the UK water industry, plus the adoption by the environmental regulators of policies to meet the requirements of legislation and public perception have reversed the trend of long-term deterioration in the urban water environment. This has been made possible by the progression of the understanding of urban wastewater pollution problems and the resultant development of associated technology and planning tools. The UPM Procedure is an example of how previously intractable problems can be addressed using logical and objective planning procedures soundly based, where appropriate, on continually improving computer modelling capabilities. These water quality models which represent the individual components of the urban wastewater system form the cornerstone to an integrated, holistic approach which is becoming more widely accepted and applied in the UK (Squibbs, 1996). Initial deficiencies in the UPM Manual (FWR, 1994) are being addressed through user experience and ongoing research and development. Similarly, perceived practical and economic constraints associated with the use of sophisticated modelling software and, in particular, the costs associated with site-specific model application data collection programmes, are gradually being overcome as the potential benefits are becoming more widely recognised through dissemination and training activities (Crabtree *et al.*, 1996b).

The future challenges lie mainly in "ground-truthing" source control technology and in wastewater engineering technology for future urban drainage schemes which are likely to be constrained by a reduction in the current scale of capital investment programmes and the need to demonstrate clear environmental cost benefits (FWR, 1996). This will result in a need for even more cost-effective solutions based on the sustainable design and operation of urban wastewater infrastructure within a framework of integrated urban pollution control. Source control and real time control technology must play a major role in achieving these goals. Also, it is likely that meeting environmental criteria will become more demanding on the wastewater system operator (Bramley, 1997). While the current organisational and policy frameworks, coupled with technology, can respond to these changing requirements, their future efficacy may be less certain if global patterns of change in climate and economic activity as well as legal challenges place greater strains on available resources and the performance of systems.

Clearly, tensions exist between levels of policy and the science and technology necessary for the rational implementation of sustainable drainage policies.

ACKNOWLEDGEMENTS

The authors acknowledge the support of Middlesex University, the Natural Environment Research Council (NERC) and the Water Research Centre (WRc) in the preparation of this material. However, the views expressed are those of the authors and not necessarily those of NERC or WRc.

REFERENCES

Argue, J.R. 1994. A new streetscape for stormwater management in Mediterranean-climate cities. *Proceedings of 17th Biennial Conference of International Association of Water Quality*, Budapest, Hungary. IAWQ, London, 23–32.

Bramley, M. 1997. Future Issues in Environmental Protection: A European Perspective. *Journal of the Chartered Institute of Water and Environmental Management*, **11** (April), 79–86.

CIRIA. 1992. *Scope for Control of Urban Runoff*. Report 124, Construction Industry Research and Information Association, London.

Clifforde, I.T. and Murrell, K.N. 1993. *Urban Pollution Management: Review of Products and Implementation*. Foundation for Water Research, Report No. FR0405.

Clifforde, I., Morris, G. and Crabtree, R. 1995. The UK Response to the Challenge of Urban Stormwater Management. *Water Science and Technology*, **32**(1), 177–183.

Colwill, D.M., Peters, C.J. and Perry, R. 1984. *Water Quality of Highway Runoff*. Report 823, Transport and Road Research Laboratory, Crowthorne.

Crabtree, R.W., Dempsey, P. and Gent, R.J. 1995. *The UPM Demonstration Project: Final Report*. Foundation for Water Research, Report No. FR0495.

Crabtree, R.W., Earp, W. and Whalley, P. 1996a. A demonstration of the benefits of integrated wastewater planning for controlling transient pollution. *Water Science and Technology*, **32**(2), 209–218.

Crabtree, R.W., Earp, W. and Whalley, P. 1996b. Data collection for UPM studies. *Proceedings of 7th International Conference on Urban Storm Drainage*, Hannover, Germany.

Department of the Environment. 1994. *Sustainable Development: The UK Strategy*. HMSO, London.

Ellis, J. B. 1986. Pollutional aspects of urban runoff. In Torno, H., Marsalek, J. and Desbordes, M. (eds), *Urban Runoff Pollution*. Springer-verlag, Berlin, 1–34.

Ellis, J.B. 1993. Achieving standards for the recreational use of urban waters. In Kay, D. and Hanbury, R. (Eds), *Recreational Water Quality Management*. Ellis Horwood, London, 155–174.

Ellis, J.B. and Hvitved-Jacobsen, T. 1996. Urban drainage impacts on receiving waters. *Journal of Hydraulic Research*, **34**(6), 771–783.

Ellis, J.B., Revitt, D.M. and Llewelyn, N. 1997. Transport and the environment: Effects of organic pollutants on water quality. *Journal of the Chartered Institute of Water and Environmental Management*, **11**(3), 170–177.

FWR. 1994. *Urban Pollution Management Manual*. Foundation for Water Research, FR/CL0002.

FWR. 1996. *Assessing the Benefits of Surface Water Quality Improvements Manual*. Foundation for Water Research, FR/CL0005.

HMSO. 1970. *Final Report on Storm Overflows*. Ministry of Housing and Local Government Technical Committee. HMSO, London.

House, M.A., Ellis, J.B., Herricks, E.E., Hvitved-Jacobsen, T., Seager, J., Lijklema, L., Aalderink, A. and Clifforde, I.T. 1993. Urban drainage: Impacts on receiving water quality. *Water Science and Technology*, **27**(12), 117–158.

Lawrence, A.I., Marsalek, J., Ellis, J.B. and Urbonas, B. 1996. Stormwater detention and BMPs. *Journal of Hydraulic Research*, **34**(6), 799–813.

LGMB (Local Government Management Board). 1993. *Local Agenda 21: A guide for local authorities in the UK*. LGMB, Luton.

Luker, M. and Montague, K. 1994. *Control of Pollution from Highway Drainage Discharges*. Report 142, Construction Industry Research and Information Association, London.

Matthews, P.J. 1995. Water quality objectives: a tool to ensure environmental protection and wise expenditure. *Water Science and Technology*, **32**(5–6), 7–14.

National Research Council. 1994. *Groundwater Recharge Using Water of Impaired Quality*. National Academic Press, Washington DC.

NRA. 1993. *Guidelines for AMP2 Periodic Review* (*Version 2*). National Rivers Authority.

NRA. 1995. *Thames 21: A planning perspective and a sustainable strategy for the Thames region.* NRA, Thames Region, Reading.

Price, M. 1994. Drainage from roads and airfields to soakaways. *Journal of Chartered Institute of Water and Environmental Management*, **8**, 468–479.

Raimbault, G. and Balades, T.D. 1987. Realisations de structures reservoirs en voirie urbaine. *Voirie Urbaine*, **64**, 39–47.

Schueler, T.R., Kumble, P.A. and Heraty, M.A. 1992. *A Current Assessment of Urban Best Management Practices.* Metropolitan Washington Council of Governments, Washington DC.

SEPA (Scottish Environmental Protection Agency). 1996. *A Guide to Surface Water Management Practices.* SEPA, Stirling.

Squibbs, G.A. 1996. Urban pollution management in the North West of England. *Proceedings of 7th International Conference on Urban Storm Drainage*, Hannover, Germany.

Tyson, J.M. 1991. Institutional issues in urban pollution management in England and Wales. *Proceedings of Engineering Foundation Conference on Urban Runoff and Receiving Waters*, Crested Butte, Colorado.

Urbonas, B., Doerfer, J.T. and Tucker, L.S. 1996. *Stormwater Sand Filtration: A Solution or a Problem.* APWA Reporter, American Public Works Association, Washington DC.

Water Research Centre. 1995. *Sewers for Adoption*, 4th edition. WRc, Swindon.

12

The Economics of Water Pollution Abatement: a Case Study

Colin Green

INTRODUCTION

The management of water and wastewater is one of the biggest problems in sustainable development. Polluting water reduces its availability for other uses; hence, the opportunity costs associated with water use can be very high in parts of the world where water is scarce (Falkenmark and Widstrand, 1992). Any use of water by removing it from the environment or returning it in a different place or a different state has such an opportunity cost, or, more strictly, two opportunity costs. The first of these costs is to the environment since the local ecosystem develops around the prevailing water balance. The second opportunity cost is to other human uses: if wastewater is discharged to the river then that river water may no longer be usable for irrigation and the costs of treatment before putting that water into potable supply may be increased. At the same time, the treatment of wastewater is typically expensive and highly capital intensive; the World Bank has estimated that there is a global spending requirement of some US$ 600 to 800 billion over the next decade. The resources so invested will not be available, for example, to improve the delivery of health care or education.

Both the Dublin Declaration (ACC/ISGWR, 1992) and Agenda 21 (United Nations, 1992) therefore lay stress upon considering the opportunity costs of the use of water. The European Union's draft Water Framework Directive (European Commission, 1997) also calls for the introduction of full resource and environmental cost pricing for all uses of water by 2010. Few economists would argue that the present low charges for most water uses result in anything which approaches the efficient use of water.

Water Quality: Processes and Policy. Edited by Stephen T. Trudgill, Des E. Walling and Bruce W. Webb.
© 1999 John Wiley & Sons Ltd.

There has consequently been a rapid expansion of the application of economic analysis to the management of water and wastewater (Green, 1996). In the water quality field, there have been analyses of the economic consequences of events such as the *Amoco Cadiz* (Bonnieux and Rainelli, 1991) and *Exxon Valdez* (Carson *et al.*, 1992) pollution incidents, and a long series of studies over the years and in several countries of the national economic costs of water pollution (e.g. Federal Ministry for the Environment, 1991) as well as studies of individual catchments (Desvouges *et al.*, 1987). These studies notably include the evaluation of those, particularly environmental, impacts which are not marketed and unpriced, and use the variety of economic techniques which have been developed in recent years for such purposes (Australian Department of the Environment, Sport and Territories, 1996). For the United Kingdom, a standard methodology with standard data for use in assessing the benefits of all projects which result in surface water quality improvements has been adopted (WRc/OXERA/FHRC, 1996).

THE ECONOMIC MODEL

Economics may be defined as the "application of reason to choice" (Green and Newsome, 1992) and economic value is an instrumental value, measuring the contribution of some action to the achievement of some objective. In neoclassical economics, the only objective considered is the maximisation of individual preferences.

It is convenient to categorise neoclassical economic value into three classes.

1. "Use" value arises from access to or consumption of a resource. Goods which are bought and sold in the market are purchased because they yield a use to the consumer. Some environmental resources also have a use value; whilst not consumed, access to the resource has a value to consumers for which they are, in principle, prepared to pay. For example, people value informal recreation by rivers or angling (WRc/OXERA/FHRC, 1996).
2. "Functional" values are a form of indirect use value (de Groot, 1987). An environmental resource may provide a function which supports a resource elsewhere where it has a use value, or the environmental resource may provide a service which would otherwise require us to commit resources to provide. For example, wetlands are well known as spawning grounds for fish and this fish population then supports a commercial or recreational fishery elsewhere (Kotze and Breen, 1994). Again, a wetland can remove certain pollutants from water or store flood flows which, in the absence of that wetland, would have to be provided by constructing a wastewater treatment works or flood alleviation scheme (Maltby, 1986).
3. "Non-use" value: economists recognise that people value environmental resources for reasons other than the direct or indirect use which can be made of them (Pearce and Turner, 1990).

Use value is thus relatively straightforward conceptually and, by now, non-priced use values are relatively routinely included into economic analyses (Ministry of Agriculture,

Fisheries and Food, 1993). Functional values present a problem in so far as determining when to include these values. The problem here is that the stocks and flows in the environment are so large relative to the flows in the economy; for example, an estimated 5 million tonnes of rain falls on Sydney and the three great rivers of Bangladesh carry some 1.5 billion tonnes of sediment. Both have a functional value, one in providing water for potable and other uses, the other in building land: should they both be evaluated? For example, Constanza *et al.* (1997) sought to compile a complete list of all the use, functional and non-use values provided by the environment.

In practice, the best answer is likely to be that the economy is conditional upon the environment. What is yielded by the economy depends upon the services provided by the environment; the relationship between the economy and environment is as a leaf to the tree. We gain nothing by seeking to evaluate the totality of the transfers from the environment to the economy since without those services there would not be an economy. What it is necessary to evaluate in any economic analysis is a change in the flows from the environment to the economy as a result of some human action. Thus, we do not, for example, need to evaluate the flood storage benefits of a wetland unless we are contemplating some action which might reduce those benefits.

Non-use value is even more problematic. Some neoclassical economists argue that whatever non-use value is, it is not a neoclassical economic value but something else (Diamond and Hausman, 1992). Some economists equally attack the attempt to incorporate such values into economics as the "commodisation" of the environment and as being morally wrong (Bilsborough, 1992) or conceptually false (Milgrom, 1992). Again, the neoclassical economic paradigm has been criticised as being conceptually inadequate and parochial (Green, 1997b). There are a variety of attempts either to extend the scope of economics, or to replace the neoclassical economic model by what its critics regard as more adequate economics, such as institutional economics, ecological economics, feminist economics or new economics.

The picture is further complicated by the varieties of neoclassical economists; some economists accept the need to supplement economic analysis by constraints, such as the "Precautionary Principle" (O'Riordan and Cameron, 1994), and Critical Natural Capital (Countryside Commission *et al.*, 1993). Some neoclassical economists further accept that there is more to value than individual preferences (Pearce and Turner, 1990).

The Evaluation of Changes in Water Quality

Economics has been defined as the "application of reason to choice" (Green and Newsome, 1992) where choice itself is necessary as a result of conflict. There is no choice to be made unless there are at least two alternatives and these alternatives are in some sense mutually exclusive. This conflict between the available alternatives can arise for a number of reasons including exclusivity in space or time, conflicting objectives and disagreements between individuals (Green, 1997a). Resources are also scarce and whilst this is not the reason why we must choose between the local alternatives, it forces us to choose whether it is worth doing anything at all and whether the best of the local alternatives is sufficiently "better" than the other alternatives to justify any additional resources required to implement it.

There are then a range of decisions with which we may be faced in terms of changes in water quality:

1. reducing the current level of human-induced pollution in the water;
2. avoiding a degradation which would otherwise occur:

   • because of a natural change;
   • as a consequence of past human activity (e.g. leachate from old landfill sites, acid rain, minewater from abandoned mines, escape of a non-indigenous plant or animal species);
   • as a risk associated with a proposed human action (e.g. the risk of failure of a silage container, a fire at a chemical plant or petrol tanker overturning on a road); or
   • as the deliberate foreseen consequence of some proposed action (e.g. to abstract water without replacement, the discharge of a polluting load).

We are faced, in each case, with two questions: whether to do anything about the pollution and, if so, who should pay the costs, although the two questions are not necessarily posed in that order.

Should Something be Done?

Neoclassical economic theory is built upon a narrow utilitarian approach, defining the single objective of individual choice as being the satisfaction of individual wants. Thus, it is, for example, inconsistent with Kant's (1785) claim that freedom resides in the power to reason about the objectives to be achieved and the selection of moral objectives as the basis of action. There is, however, increasing evidence that people bring moral values to choices involving the environment (Green and Tunstall, 1996) and Douglas (1966) argued that the label of "pollution" defines an issue as a moral wrong. It also takes individual choice as a general theory of choice so that collective choices can be determined by some aggregate individual wants. This, it has been argued (Green, 1997a), is to make two sweeping presumptions: that people both approach collective choices as if they were individual choices and also expect collective choices to be made as if they were individual choices.

The neoclassical economic rationale for doing something about pollution is the Pigovian externalities model where an externality is a consequence of one person's action, to which no price is attached, that is borne by another person. Such externalities may either increase the resource requirements for production, or reduce the value of consumption yielded by that other person's activities. Since the externalities have a zero cost to their producer, it is anticipated that these externalities are being produced in excessive amounts and if instead they were priced at the value of their impact upon the recipient then economic efficiency would be enhanced. In the Pigovian model, these externalities should be internalised into the decision of the person causing those externalities through a tax or charge. As an approach, it says nothing about whether the producer should be charged or the recipient required to pay to reduce those externalities; for example, whether a noisy factory should be required to compensate its neighbours for its noise or whether those neighbours should have to pay the factory to fit noise suppression equipment.

There are a number of practical problems with this model. It assumes, for example, that the correct economic valuations of those externalities are both knowable and known: in the case of pollution, that we can determine accurately all of the use, functional and non-use values associated with each and every impact. Setting efficient levels of tax requires that the abatement costs also be known. The information requirements are consequently very large; thus, the implementation of full cost pricing of all uses of water proposed in the draft EU Water Framework Directive (European Commission, 1997) requires both extensive river flow gauging and water quality monitoring on all watercourses as well as a sufficient length of record so that extreme flows can be accurately predicted. This is an optimistic assumption.

A more general problem with the Pigovian approach is that it would be easy were a decision to result in only one externality and that there were only one decision. If a decision results in multiple externalities, as is the case with much development, there is a risk that introducing a charge for only one of these externalities will cause further distortions, a problem known as the "second best" problem (Lipsey and Lancaster, 1956–57). Where the situation is already far from optimal, introducing an optimal approach to only part of the problem can move the situation further away from the hypothetical optimum. For example, a tax on the development of floodprone land will not result in an overall gain if the net effect is simply to shift development to nearby land which has a high landscape or environmental value (Green *et al.*, 1996). Again, a charge for wastewater alone might not result in the Best Environmental Option for the disposal of waste being adopted.

Finally, there is widespread evidence that industrial demand for water and wastewater is inefficient (Porter and van der Linde, 1995; Rees *et al.*, 1993); thus, although companies are paying for water used and wastes discharged, they could increase their profitability by reducing consumption and discharges. Porter and van der Linde (1995) have pointed out that pollution is also a waste since it involves throwing away resources. For example, wastewater is water, which has often been treated at some cost and bought in from some water supplier; it may have a significant residual heat load, and typically includes in suspension or dilution materials used either as raw materials to the production process or as part of that process. Indeed, it is commonly found that when the existing practices of industrial companies are examined, they can significantly reduce pollution and increase profitability at the same time (Hills, 1995). In such cases, pollution abatement does not cost money but saves it; 15–25% improvements at zero cost are not uncommon (Greer and van Loben Sels, 1997; Hills, 1995; Johnston, 1994) with some authors arguing that, over time, much greater reductions will be achieved without significant additional costs (Lovins *et al.*, 1997).

Who Ought to Pay?

Neoclassical economics has sought desperately to make claim to be a value-free "science" and consequently has attempted to avoid entering any ethical or moral debate. For example, in neoclassical economics, the rich and vexed question of equity is redefined to refer solely to the distribution of income. In practice, the axioms of neoclassical economics, notably the assertion that value is solely determined by individual preference, make very strong moral claims. One way that neoclassical economics has sought to avoid such involvement in moral questions is by relying upon the claim that there is a "natural" allocation of rights, this allocation of entitlements and obligations being termed "property rights". By first redefining the question away from "what is right" to "who has the right", then attaching this right to the possession of property, and finally asserting that such rights are natural, the original question can be evaded. Furthermore, by assuming that rights are natural then these rights can also be taken to be permanent and not subject to revision. This claim is reinforced by libertarian claims that individual property rights are the highest form of liberty. Conversely, if individual entitlements and obligations are taken to be socially determined then they cannot be taken to be fixed nor can the question be evaded of what they should be. The neoclassical economic approach also supposes a generality to the Anglo-Saxon approach to property rights which does not exist; consequently, neoclassical economics can be argued to be parochial (Green, 1997b).

The second attempt to escape from the necessity to determine "what is right" was by Coase (1960). Coase sought to demonstrate that provided there was some given distribution of property rights, it did not matter what that distribution was; efficiency could be achieved in any case. Thus, it does not matter whether a factory has the right to emit noise or its neighbours have a right to peace and quiet since an efficient solution can be achieved however rights are allocated. Unfortunately, such a claim can only be maintained if the costs of reaching an agreement are ignored. Equally, it is an interesting assumption that people do not care about the distribution of entitlements and obligations and that their sole objective is economic efficiency.

What Coase's model does require is some fixed allocation of entitlements and obligations, and neoclassical economists complain if property rights are not clear. In practice, a fundamental problem with the creation of private property rights is often the very virtue claimed for it by neoclassical economists. Being fixed, it is inflexible in terms of efficiency. Thus, for example, in those countries which have adopted the colonial Anglo-Saxon model of creating private property water rights to water abstraction, two common problems are: an initial allocation of more rights than is sustainable and an allocation which now no longer results in that water being put to its highest and best use. Whilst an elegant system for colonising lands assumed to be both empty and infinite in resource, it is clumsy and inefficient when the problem is one of managing a fixed resource under pressure and under changing conditions. More generally, the success of a natural rights approach requires that the system will be able to cope with any future change in conditions and that allocated rights will never conflict.

The particular problem with water rights is that the resource is exogenously determined by events, including climate change, and actions perhaps hundreds of kilometres away. From a hydrological perspective, a more logical approach would have been the allocation of rights to runoff, rather than to abstraction, but adopting such an approach would now be likely to create more problems than it solves. Again, fixed rights presume an unchanging social allocation of entitlements and obligations whereas in reality these evolve and change over time. In the nineteenth century, it was presumed that employees accepted all of the risks of employment as part of their contract and that purchasers were fully aware of all the risks when buying a good. This view has now largely been reversed and a similar shift with respect to the environment also appears to be taking place (Ingram and Oggins, 1992).

One view of the current interest in the use of economic instruments, tradable permits (Merrett, 1997) and environmental charges (Carlin, 1992; European Environment Agency, 1996) is then that they are expedients to dealing with the failings of Anglo-Saxon law, the former enabling the reallocation of existing rights to higher and better uses and the latter reducing demand to the sustainable yields. This is not to say that economic instruments are not useful, and perhaps ethically desirable, but only that too much should not be claimed for them.

Thus, the neoclassical economic model provides no escape from the question of who ought to pay. An alternative approach then is to determine what are the objectives, including both efficiency and equity, that it is sought to achieve and the nature of the problem that must be managed. South Africa, faced with a peculiarly awkward mixture of Anglo-Saxon law and Romano-Dutch law, which embedded the inequities of the previous regime, is setting out to establish a new basic water law from first principles, establishing what are the principles and requirements that such a law should achieve (Department of Water Affairs and Forestry, 1996). Unfortunately, this is not an option which is open to all countries.

To Pollute or Not to Pollute

The neoclassical economic model defines the sole objective of collective choice as being the maximisation of individual wants with respect to available resources: efficiency. Irrespective of whether the issue is one of reducing the current level of pollution, or examining the potential for increased pollution, the neoclassical model formulates the problem in exactly the same way. The benefits of reducing pollution, or conversely the disbenefits of increased pollution, are compared to costs of pollution abatement. At the point where the marginal costs of pollution abatement equal, as appropriate, either the marginal benefits of pollution abatement or the marginal disbenefits of increased pollution, here is the efficient, or optimal, level of pollution. Whilst economists use the term "efficient" rather than "optimal", in economic terms "efficient" has the meaning of optimal. The neoclassical economist approaches a pollution question with the expectation that some non-zero amount of pollution will be the ideal solution; indeed, sometimes it is concluded that there should be more pollution rather than less.

Thus, the neoclassical economic model is a very limited conceptualisation of collective choices, embodying a series of rather debatable assumptions. Unfortunately, the neoclassical paradigm is the only developed form of economic analysis and, if we seek to apply reason to choice, then it is necessary to adopt a neoclassical economic analysis as the starting point. The decision of how to manage minewater discharges from Grootvlei illustrates these problems in the real world and also the lesson that the aim of any form of analysis is to get a better understanding of what is involved in the decision and not simply to get some numbers.

THE PROBLEM

The East Rand Basin, east of Johannesburg, is a declining area for gold mining; the number of mines fell from some 22 large mines in the 1950s to five in the 1980s and those were marginally profitable. Gold mines are also deep and require extensive pumping in order to remain workable. Since the mines are interconnected for safety reasons, it is possible to use one pumping station to drain the entire basin. This pumping station is, logically, located below the deepest level of the deepest working mine. In the mid-1980s, this mine was the Sallies mine and the costs of pumping were shared by all the operating mines in the basin; the State also subsidised the pumping costs.

With the closure of the Sallies mine, it was considered that if the pumping instead took place from Grootvlei No. 3 shaft then the State could reduce the subsidies required by a total of around R64 million. After the necessary permit, which specified the pollutant load that could be discharged, was granted by the Department of Water Affairs and Forestry (DWAF), pumping operations commenced at the No. 3 shaft in November 1995. Unfortunately, both the volume and quality of the water pumped were worse than predicted and breached the permitted limits. Moreover, the No. 3 shaft and the discharge point for the untreated water were immediately adjacent to the Blesbokspruit, a Ramsar designated wetland. The immediate result was a red plume of water spreading downstream through the wetland as iron and manganese oxidised and were then deposited on the bed of the wetlands. This particularly affected bottom-feeding fish and the death of these fish reduced the food supply for some bird species and hence the number of those species using the wetlands. The DWAF withdrew the discharge permit and required the installation of treatment facilities before an interim permit could be issued. A lagoon system for aeration, liming and clarification was installed, pending the completion of permanent works; an interim permit was issued and pumping recommenced.

Minewater is commonly loaded with heavy metals and is acidic; both these problems are managed by the aeration, liming and clarification plant installed. Unfortunately, the minewater is also very saline and this salinity poses a number of threats to downstream uses; but this salinity can only be reduced by partial or complete desalination, a process which is expensive both in capital and operating terms.

The Government was therefore faced with a severe policy problem, with the final decision being reserved to the Cabinet. All the apparent options were markedly unattractive.

1. If the pumping were to be stopped, then 6000 jobs would be lost in an area where unemployment is already 30% and when the South African Government has given high priority to expanding the economy so as to create jobs.
2. The minewater has been damaging a wetlands designated under the Ramsar Treaty and it was argued that should the South African Government allow such damage to occur, it would reduce the credibility of South Africa's commitment to the environment, a right guaranteed by the Constitution, and reduce the confidence of foreign investors where foreign investment is in turn essential if the economy is to be expanded.

3. The Constitution guarantees environmental rights; politically, if the Government were to be seen to violate one constitutional right then its credibility with respect to protecting all constitutional rights would be reduced.
4. The Government has committed itself to an emergency water and sanitation programme; the scale of this problem is shown by the initial objective of providing a water point within 200 m of every household in the 12 000 to 15 000 communities currently lacking such facilities (Department of Water Affairs and Forestry, 1994). Any Government-financed treatment of the minewater could be at the expense of the progress of the emergency water and sanitation programme.

The Government therefore set up an inter-departmental committee, the Grootvlei Joint Venture Committee, to commission research on the issues and to undertake a cost–benefit analysis relating to mining in the East Rand Basin (Grootvlei Joint Venture Committee, 1996). As part of this process, the Water Research Commission asked the Foundation for Water Research to apply the UK methodology to the assessment of the options (Foundation for Water Research, 1996). The following discussion draws heavily upon the latter report but with the benefit of hindsight.

The Options

Any cost–benefit analysis is a comparison of a series of options against some baseline option. This, depending on the circumstances, is the "do nothing" option, the "do minimum" or "do the same as now" option. Since the permanent aeration, liming and clarification works was under construction, the baseline option in this case was taken to be "do the same as now". The other options branched out tree-like but the two main branches were: to stop pumping and close the mines, or to install one or another form of desalination and dispose of the treated water either to the Blesbokspruit or to potable supply. If pumping is stopped then the water will gradually rise and after five to seven years it was anticipated that it would start to decant downstream of the wetlands at Nigel. Figure 12.1 is a tree diagram which sets out the alternative options which were compared against the baseline option.

There are suboptions within each of these options: for example, the means of disposing of the iron sludge, and the technologies to adopt for desalination and brine disposal. In order to contain the number of options compared within manageable limits, judgements have been made as to the most appropriate technologies to adopt in each case. Again, eventually the mines will close, and a decision has had to be taken as to what option will then be adopted (e.g. to continue pumping but from a higher level or to allow water to decant at Nigel).

The Benefits

A wide range of impacts from one or another option were identified. Merrett (1997) reports the problems of trying to make an estimate of the affordability of water charges in the absence of information on income. This is perhaps a particularly extreme case of

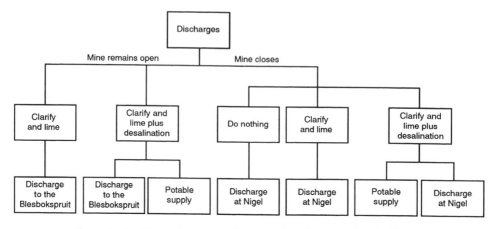

Figure 12.1   *The main options for managing the Grootvlei discharges*

the general problem of applied economic analyses: the lack of data or the existence of conflicting data. For example, although wetlands are well known to yield benefits in terms of removing polluting loads (Maltby, 1986), and artificial wetlands are constructed for that purpose (Crites, 1994), there were insufficient water quality sampling data to determine what pollution load is removed by the wetlands at present. Consequently, this benefit could not be evaluated. Similarly, quite large differences were found in the estimates of the quantities of water used for irrigation and area under irrigation which made problematic the estimation of the value of agricultural product which would be lost if irrigation could no longer take place. Other categories of benefit which were evaluated, such as the recreation value of the wetlands, proved to have trivial benefits. Therefore, the following discussion covers only the five most important categories of benefit to the decision: mining, potable water, irrigation, Critical Natural Capital and Constant Natural Asset, and the dolomitic layer. Furthermore, a structure to the choice is given which only became apparent in hindsight.

The choice of the option depended first of all on the economic benefits of continued mining. If these were marginal, and the annual profits of the mines in the area were relatively small, then closure of the mines would be likely to be the best option. If the economic benefits of mining were sufficient to justify their continued operation then choice of the option would depend upon the cost of further treatment versus the benefits and avoided damage resulting from that treatment.

Mining

The very high unemployment rate indicated that the opportunity costs of the labour employed in the mine were well below the financial costs of wages. Methods of estimating the opportunity cost, or shadow wage, of labour are well established (Squire and van der Tak, 1975) if difficult to apply in practice. Using available data, two different estimates of this opportunity cost were made. The estimates of the yield from mining were further adjusted to take account of the remittance of wages and profits overseas. The net effect was to show that the economic value of the mines was significant, with a present value of around R580 million.

There was evidence (Central Economic Advisory Service, 1989) that, in addition to the contribution by the Government to the costs of pumping which was already included in the analysis, electricity prices had also been subsidised in the past. Lack of time prevented the appropriate adjustment being made for current practice.

As export earners, the contributions of the mines to the South African economy depend critically upon both the level of gold prices and the dollar–rand exchange rate in the future. So, too, do the proportions of gold reserves which it will be financially viable to extract: the higher the gold price and the higher the dollar–rand exchange rate, then the greater the contribution of the mines to the economy, the longer the viable life of the mines and the lower the grade of ore which it will be viable to mine. The assumption was made that neither the exchange rate nor the world price for gold would change in real terms in the future. Therefore, whilst the mine has significant reserves of low grade ore, it was assumed that it would be uneconomic to mine these and that the economic life of the mine would be a minimum of 12 years and a maximum of 30 years at current output levels. Estimates of the total reserves of gold are somewhat misleading as some seams are likely to be too low grade to be recoverable unless the dollar price of gold greatly increases or costs of mining are greatly reduced. In this particular instance, the final conclusion did not depend upon future movements in the world price of gold and dollar–rand exchange rate.

Critical Natural Capital or Constant Natural Asset

The Blesbokspruit wetlands are designated under the Ramsar Convention. Thus, two critical questions were: the extent to which discharges would damage the wetlands; and how to incorporate that damage potential into the cost–benefit analysis in a way which would both work and help the decision process. Whilst some economists have sought to attach a monetary value to the loss of sites (Hanley and Spash, 1993), including "non-use" or "passive use" value, a number of objections have been put forward to such a procedure. Some environmentalists reject outright the attempt to put a money value on environmental damage (Foster, 1997); thus, in terms of promoting a collective understanding of the decision, any attempt to put a money value on the potential damage might work but certainly would not help. Again, some studies (Burgess *et al.*, 1997; Schkade and Payne, 1994; Vadnjal and O'Connor, 1994) have shown that people do not approach such choices in the way presumed by neoclassical economics. Neoclassical economists themselves, it was noted earlier, do not agree whether non-use value is part of neoclassical economic value, or whether it can be measured reliably. Neoclassical economic theory has also been argued to provide an inadequate theoretical framework for the analysis of such choices (Green, 1997b). Finally, the neoclassical economic system does not incorporate the concept of "inherent value", that species have an inherent value by right of existence (Naess, 1993), and this would have to be included into the analysis in other ways. In these circumstances, to seek to derive non-use values was considered to be neither theoretically sound nor helpful.

However, since the effects upon the wetlands, and the relative importance which should be given to any damage that is likely to occur as a result of the discharges, is crucial to the choice that must be made, it was not possible to simply opt out of the problem of incorporating these effects into the calculus of choice. The approach adopted was the use of the concepts of Constant Natural Assets and Critical Natural Capital (Countryside Commission *et al.*, 1993). For any category of environmental asset deemed to be the former, provided that the aggregate total of such assets remains constant, it is acceptable for one such asset to be lost. In effect, any loss must be replaced by an equivalent site. Such a conceptualisation may be argued to underlie the "No Net Loss" approach adopted in the United States with respect to wetlands (Heimlich, 1991) and its implementation through "habitat banking" (Reppert, 1992). Clearly, for such an approach to be feasible, it must be possible to create an area of habitat which is in some sense equivalent to that which is being lost. Moreover, this equivalent area, as exemplified in the "habitat banking" approach, should have been created before the loss of the area for it is to be a substitute.

Environmental resources which are designated as being part of Critical Natural Capital may not be sacrificed, because they have no reasonable substitute, except in extreme circumstances. The European Union's Habitat Directive (92/43) may be argued to embody the Critical Natural Capital approach by allowing loss or damage only where there is no alternative option *and* there are "imperative reasons of overriding public interest" why the works causing the damage must be undertaken.

Consequently, if the Blesbokspruit reserves, a designated Ramsar site, is part of Critical Natural Capital then there can be no question of allowing any discharge which would result in damage to that site. Conversely, if it is better classed as part of Constant Natural Assets then any damage would be acceptable providing that equivalent sites were created or provided. Strictly, such sites should have been created prior to the loss occurring but this could, in this instance, be counterbalanced by requiring a greater area to be provided in compensation than the area lost.

The Ramsar Convention allows for the treatment of Ramsar sites as Constant Natural Assets (Article 4, paragraph 2) but this does not imply that a Ramsar site may not be part of Critical Natural Capital. In this instance, the evidence showed that the existing wetlands were themselves created by relatively recent human activity, the road, rail and pipeline embankments partly damming the river corridor and resulting in permanent wetlands. The peat samples taken in the wetlands (Breen *et al.*, 1996) were all, except one at the very top of the wetlands, dated after the start of nuclear testing, giving the date of earliest peat deposition to be slightly over 50 years ago. It was also found that some 50% of the water flow in the river originates from the discharges of the wastewater plants upstream; given that the potable water supplies are pumped up from the Vaal Barrage rather than originating in the catchment, the seasonal stability of the flow in the wetlands and a large part of the total flow may also be argued to result from human intervention. On these grounds, it was concluded that it would be appropriate to treat the wetlands as a Constant Natural Asset rather than as part of Critical Natural Capital.

The deposition of metal salts undoubtedly caused major damage to the wetlands but this deposition will stop with the completion of the permanent aeration, liming and clarification works. The wetlands will, however, continue to be subject to both a high salt load and also to increased water flows. We did not find an ecologist who was prepared to argue that the saline load would damage the value of the wetlands; indeed, some argued that the increased salinity would make the wetlands more attractive to some Red List species. In the end, it was concluded that the main problem would be the increased flow of water, which would double current flows, as there is a feeling that the wetland is already suffering from overwatering. However, the policy of extending basic potable water and sewerage provision to the population necessarily means that wastewater flows to rivers will increase and the major proportion of the current flow comes from the upstream treatment works. Therefore, the local sewerage utility was asked to provide estimates of the growth to be expected in those discharges; their estimate was that growth in those discharges over the next 12 years would equal the current flow in the river. Thus, if there were no minewater discharges then flows through the wetlands will double and if minewater is discharged then they will treble. Therefore, it was recommended that a management plan be developed for the wetlands, including a water management plan, and that the ecological health of the wetlands should be monitored. As a result of that monitoring, it may become necessary either to provide a bypass channel for the main flow through the wetlands or to construct an equivalent site.

Irrigation

If sufficient other harm were to be caused by the minewater then it would still be appropriate either to close the mines or to invest in further treatment before discharge. The high salinity of the discharges will make the river waters unsuitable for irrigation use and hence those farms downstream which currently use irrigation will have to abandon its use.

Irrigation yields benefits in one or more of the following four ways (Weatherhead *et al.*, 1994):

- by reducing the risk of crop loss as a result of a drought;
- by allowing an increase in the yield of crop grown;
- by allowing a shift to a higher valued crop but one which is more sensitive to water availability; or
- by ensuring the quality of the crop grown so that a quality premium is gained.

On the other hand, water which is used for irrigation is largely lost through evaporespiration. Therefore, water which is freed from irrigation use is available for other uses which may have a higher value than its use for irrigation and which, like most potable water uses, return a far higher fraction of the water used than do irrigation uses.

In order to evaluate the value of the water currently used for irrigation, and to compare it to its value in other uses, estimates are required of:

- how much land is being irrigated;
- by how much water;
- for what crops;

- what cropping pattern would be followed if irrigation water were not available; and
- the relative profitability of the farming under the cropping pattern with irrigation water compared to without water.

The estimates of the area under irrigation varied between 640 and 1000 ha. The amounts of water that the farmers abstract is not metered and three widely differing estimates of the amounts of water which are abstracted were also found:

- 16 557 ml/annum based upon an assumed rate of application of irrigation water;
- 3020 ml/annum based upon the amount of water which disappears from the system; and
- 4808 ml/annum (Breen *et al.*, 1996).

Breen *et al.* (1996) based their assessment on advice from the Bureau of Statistical and Survey Methodology at the University of Pretoria as to both present crop mixes and dryland farming, and the gross margins associated with each cropping pattern. This assumes that all of the land would be converted to maize growing. The resulting reduction is R2 486 000/year, very small relative to the cost of desalination.

Potable Water Supplies

Thus, the economic justification for any desalination had largely to lie in the benefits from increasing the local water resource. Discharge of minewater will increase the salinity problem which is already affecting the Vaal Barrage, the principal water resource for the province. Conversely, desalination could yield an additional resource close to the centre of demand. However, the new resource would equal a maximum of 2% of current output whilst demand is growing at approximately 3.5% annually in Gauteng.

   The local water utility, Rand Water, estimates the marginal value of potable water available in the Springs area as being around 55 cents/m^3, this being the saving in pumping treated water up to the area from their existing works. Thus, the benefit of desalination would be in the order of R11.7 million/year. That water which was not lost through leakage and evapo-transpiration and so was returned to the river after wastewater treatment would be available for reuse but no value was put on this water. These benefits were considerably below the estimated capital and operating costs of a desalination plant, which had a present value of R3000 million, the other benefits resulting from desalination being insufficient to justify desalination.

Dolomitic Resource

An outstanding question at the beginning of the Joint Venture Committee study was the consequences of ceasing pumping and allowing water in the basin to rise until it naturally discharged at Nigel. The basin is overlain by a layer of dolomite. Scott (1995) estimated that the total inflow to the dolomite layer from rainfall might be of the order of 51 Ml/day. This would be the upper bound of the potential yield from the aquifer and could be evaluated in the same way as above. Since this water is currently a significant part of the flows which require to be pumped from the mines, such use would reduce the pumping costs, where again this reduction can be evaluated both in terms of the savings of pumping and treatment costs and also in terms of the down-stream consequences of discharging the water from Grootvlei.

In addition, both the Scott (1995) report and an unpublished report by the Council for Geoscience concluded that, if the water level were to be allowed to rebound to its natural level, then there would be a risk of polluted water breaking through the surface as springs, so spreading pollution, thus, there would be a risk of dispersed and diffuse pollution. Furthermore, it is likely that additional sinkholes might be formed. Neither loss can be evaluated without more data but the potential consequences are such as make the acceptability of this option appear very problematic. One consequence is then that pumping from some depth will have to continue after the mines have closed.

LESSONS

The final conclusion of the analysis was that neither closing the mines nor the use of desalination was economically justified, and that the existing approach of aeration, liming and clarification was the best option. Two almost equally important findings were, first, that the wetlands were probably under threat irrespective of the decision about the minewater discharges because of the anticipated doubling of flows through the wetlands from the wastewater plants upstream. Secondly, the conclusion that the uncertainties as to the consequences of allowing the water in the basin to rebound to its natural level were too great for this to be an acceptable option. Taken together, the lesson to be drawn from this study is that it is the analysis of the consequences of the options, which can in turn lead to the identification of new options, which is important rather than the monetary values which are derived.

Methodologically, the lesson is to look at what is happening in the surrounding area since it was only in this way that the anticipated increase in discharges from the upstream wastewater treatment works was discovered. As ever in real cost–benefit analyses, the data available were sparse and conflicting; however, the analysis showed that the conclusions were quite robust to these uncertainties.

REFERENCES

ACC/ISGWR. 1992. *The Dublin Statement and the Report of the Conference*. World Meteoro-
logical Organization, Geneva.
Australian Department of the Environment, Sport and Territories. 1996. *Techniques to Value
Environmental Resources: an Introductory Handbook* (http://www.erin.gov.au/portfolio/dest/
eeu/est//estim1.htm).
Bilsborough, S. 1992. The oven-ready golden eagle: arguments against valuation. *ECOS*, **13**(1),
46–50.
Bonnieux, F. and Rainelli, P. 1991. *Catastrophe Ecologique & Dommages Economiques*. INRA/
Economica, Paris.

Breen *et al.* 1996. *The costs to the environment of pumping minewater into the Blesbokspruit.* Report to the Department of Environmental Affairs and Tourism, South Africa.

Burgess, J., Clark, J. and Harrison, C.M. 1997. '*I struggled with this money business': respondents to the wildlife enhancement scheme CV survey discuss the validity of their WTP figures.* School of Geography, University College, London.

Carlin, A. 1992. *The United States Experience with Economic Incentives to control Environmental Pollution.* EPA-230-R-92-001, US Environmental Protection Agency, Washington, DC.

Carson, R.T., Mitchell, R.C., Hanemann, W.M., Kopp, R.J., Presser, S. and Ruud, P.A. 1992. *A Contingent Valuation Study of Lost Passive Use Values Resulting from the Exxon Valdez Oil Spill.* A Report to the Attorney General of the State of Alaska.

Central Economic Advisory Service. 1989. *Manual for Cost-Benefit Analysis in South Africa.* Central Economic Advisory Service, Pretoria.

Coase, R.H. 1960. The Problem of Social Cost. *Journal of Law and Economics*, **3**, 1–44.

Constanza, R., d'Arge, R., de Groot, R., Farber, S., Grasso, M., Hannon, B., Limburg, K., Naeem, S., O'Neill, R.V., Paruelo, J., Raskin, R.G., Sutton, P. and van den Belt, M. 1997. The value of the world's ecosystem services and natural capital. *Nature*, **387** (15th May), 253–260.

Countryside Commission/English Heritage/English Nature. 1993. *Conservation Issues in Strategic Plans.* Countryside Commission, Northampton.

Crites, R.W. 1994. Design criteria and practice for constructed wetlands. *Water Science and Technology*, **29**, 1–6.

de Groot, R.S. 1987. Environmental function as a unifying concept for ecology and economics. *Environmentalist*, **7**(2), 105–109.

Department of Water Affairs and Forestry. 1994. *Water Supply and Sanitation Policy: Water – an indivisible national asset.* White Paper WP-1, Republic of South Africa, Cape Town.

Department of Water Affairs and Forestry. 1996. *Water Law Principles.* Department of Water Affairs and Forestry, Pretoria.

Desvouges *et al.* 1987. Option price estimates for water quality estimates. *Journal of Environmental Economics and Management*, **14**, 248–267.

Diamond, P.A. and Hausman, J.A. 1992. On contingent valuation measurement of nonuse values. In *Contingent Valuation: A Critical Assessment.* Cambridge Economics Inc., Cambridge, MA.

Douglas, M. 1966. *Purity and Danger.* Routledge, London.

European Commission. 1997. *Commission Proposal for a Council Directive establishing a Framework for Community Action in the Field of Water Policy.* European Commission, Brussels.

European Environment Agency. 1996. *Environmental Taxes: Implementation and Environmental Effectiveness.* Environmental Issues Series 1, European Environment Agency, Copenhagen.

Falkenmark, M. and Widstrand, C. 1992. Population and water resources: a delicate balance. *Population Bulletin.* Population Reference Bureau, Washington, DC.

Federal Ministry for the Environment. 1991. *Advantages of Environmental Protection; Costs of Environmental Pollution.* Bonn.

Foster, J. (Ed.) 1997. *Valuing Nature? Economics, Ethics and Environment.* Routledge, London.

Foundation for Water Research. 1996. *Grootvlei Socio-Economic and Environmental Cost Benefit Analysis.* Report to the Water Research Commission, South Africa, Foundation for Water Research, Marlow.

Green, C.H. 1996. Economic analysis: what has been learnt so far? In Howsam, P. and Carter, R.C. (Eds), *Water Policy: Allocation and Management in Practice.* Spon, London.

Green, C.H. 1997a. Are Blue Whales really simply very large cups of coffee? Paper given at the *7th Stockholm Water Symposium.*

Green, C.H. 1997b. Water, the environment and economics: what does experience teach us so far? *Canadian Water Resources Journal*, **22**(1), 85–97.

Green, C.H. and Newsome, D. 1992. Ethics and the Calculi of Choice. *A Holistic Approach to Water Quality Management – Proceedings of the Stockholm Water Symposium.* Stockholm Vatten, Stockholm.

Green, C.H. and Tunstall, S.M. 1996. The environmental value and attractiveness of river corridors. In Bravard, J.-P., Laurent, A.-M., Davallon, J. and Bethemont, J. (Eds), *Les paysages de l'eau aux portes de la ville*. Centre Jacques Cartier, Lyon.

Green, C.H., van der Veen, A., Reitano, B., Wierstra, E., Ketteridge, A.-M., Otter, H. and Rivilla, M. 1996. *The Use of Economic Instruments in Catchment Management*. EUROFLOOD Technical Annex, Report to the European Commission, Flood Hazard Research Centre, Enfield.

Greer, L. and van Loben Sels, C. 1997. When pollution prevention meets the bottom line. *Environmental Science and Technology* (http://pubs.acs.org/hotartcl/est/97/sept/when.html).

Grootvlei Joint Venture Committee. 1996. *Report of the Grootvlei Joint Venture Committee in Respect of a Preliminary Cost-Benefit Analysis should the Pumping of Mine Water into the Blesbokspruit Continue*. The Committee, Johannesburg.

Hanley, N. and Spash, C.L. 1993. *Cost-Benefit Analysis and the Environment*. Edward Elgar, Aldershot.

Heimlich, R.E. 1991. *National Policy of 'No Net Loss' of Wetlands: What Do Agricultural Economists Have to Contribute?* Resources and Technology Division, US Department of Agriculture, Washington, DC.

Hills, J.S. 1995. *Cutting Water and Effluent Costs*. Institution of Chemical Engineers, London.

Ingram, H. and Oggins, C.R. 1992. The public trust doctrine and community values in water. *Natural Resources Journal*, **32**, 515–537.

Johnston, N. 1994. *Waste Minimisation: a route to profit and cleaner production – an interim report on the Aire and Calder project*. Centre for Exploitation of Science and Technology, London.

Kant, I. 1785. *Fundamental Principles of the Metaphysics of Morals* (translated by T.K. Abbott) (gopher://gopher.vt.edu: 10010/02/10715).

Kotze, D.C. and Breen, C.M. 1994. *Agricultural Land-use Impacts on Wetland Functional Impacts*. Report to the Water Research Commission, WRC Report No 501/3/96, Institute of Natural Resources and Department of Grassland Science, University of Natal.

Lipsey, R.G. and Lancaster, K. 1956–57. The general theory of second best. *Review of Economic Studies*, **24**, 11–32.

Lovins, A., Lovins, H. and von Wiezsacker, E. 1997. *Factor Four*, Earthscan, London.

Maltby, E. 1986. *Waterlogged Wealth*. Earthscan, London.

Merrett, S. 1997. *An Introduction to Hydroeconomics: Political Economy, Water and the Environment*. UCL Press, London.

Milgrom. P. 1992. Is sympathy an economic value? Philosophy, economics and the contingent valuation method. In *Contingent Valuation: A Critical Assessment*. Cambridge Economics Inc., Cambridge, Mass.

Ministry of Agriculture, Fisheries and Food. 1993. *Flood and Coastal Defence: Project Appraisal Guidance Notes*. HMSO, London.

Naess, A. 1993. The deep ecological movement: some philosophical aspects. In Armstrong, S.J. and Botzler, R.G. (Eds), *Environmental Ethics – Divergence and Convergence*. McGraw-Hill, New York.

O'Riordan, T. and Cameron, J. (Eds) 1994. *Interpreting the Precautionary Principle*. Earthscan, London.

Pearce, D.W. and Turner, R.K. 1990. *Economic of Natural Resources and the Environment*. Harvester Wheatsheaf, Hemel Hempstead.

Porter, M.E. and van der Linde, C. 1995. Toward a new conception of the environment–competitiveness relationship. *Journal of Economic Perspectives*, **9**(4), 97–131.

Rees, J.A., Williams, S., Atkins, J.P., Hammond, C.J. and Trotter, S.D. 1993. *Economics of Water Resource Management*. R & D Note 128, Foundation for Water Research, Marlow.

Reppert, R. 1992. *National Wetland Mitigation Banking Study*. IWR Report 92-WMB-1, Institute for Water Resources, US Army Corps of Engineers, Alexandria, VA.

Schkade, D.A. and Payne, J.W. 1994. How people respond to contingent valuation questions: a verbal protocol analysis of willingness to pay for an environmental regulation. *Journal of Environmental Economics and Management*, **26**, 71–89.

Scott, R. 1995. *Flooding of Central and East Rand Gold Mines: An investigation into controls over the inflow rate, water quality and its predicted impacts of flooded mines.* Report to the Water Research Commission, WRC Report No. 4861/95, Institute for Groundwater Studies, University of the Orange Free State.

Squire, L. and van der Tak, H. 1975. *Economic Analysis of Projects.* Johns Hopkins, Baltimore.

United Nations. 1992. *Agenda 21: The United Nations Programme of Action From Rio.* United Nations, New York.

Vadnjal, D. and O'Connor, M. 1994. What is the value of Rangitoto Island? *Environmental Values,* **3**, 369–380.

Weatherhead, E.K., Price, A.J., Morris, J. and Burton, M. 1994. *Demand for Irrigation Water.* National Rivers Authority, R & D Report 14, HMSO, London.

WRc/OXERA/FHRC. 1996. *Assessing the Benefits of Surface Water Quality Improvements: Manual.* Foundation for Water Research, Marlow.

13

Inorganic Farm Wastes: an Environmental Time Bomb?

Ian Foster, Mark Hancock and Brian Ilbery

AGRICULTURAL CHANGE AND FARM WASTE

Waste has been defined as substances or objects which fall out of the commercial cycle or out of the chain of utility; as substances which are discarded, disposed of or got rid of by their holder; and as substances which can only be used after being subjected to a special recovery operation (Waste Management Licensing Regulations, 1994). Estimates of the amount of waste produced annually in the UK vary considerably, from 2500 million tonnes (House of Commons Environment Committee, 1989) to 500 million tonnes (RCEP, 1985). Whilst waste and pollutant discharge are regulated in every major UK industry, the effectiveness of this regulation, and compliance with legal requirements, are frequently questioned in rural areas because agriculture is exempt from certain legislation, and regulations are geared to agriculture and not the environment (Hawkins, 1984; Cox *et al.*, 1986; Munton *et al.*, 1990; Lowe *et al.*, 1992).

The environmental disbenefits of "productivist" agriculture are well known and alternative discourses associated with environmental protection, sustainability and food quality have now arisen (Bowers, 1995). Nevertheless, there remains relatively little research on farm waste, even though it is estimated that 200 million tonnes are produced annually. Nearly two-thirds of this comes from cereal straw and most of the rest is organic livestock wastes such as slurry and manure, totalling approximately 78 million tonnes per annum (Ministry of Agriculture, Fisheries and Food (MAFF), pers. comm.). The potential impact of liquid agricultural wastes on aquatic systems is well

Water Quality: Processes and Policy. Edited by Stephen T. Trudgill, Des E. Walling and Bruce W. Webb.

documented. Milk, silage effluent, pig slurry and cattle slurry have typical BODs (biochemical oxygen demand) of 140 000, 30 000–80 000, 20 000–30 000 and 10 000–20 000 mg/l respectively, in comparison with treated domestic sewage which is 20–60 mg/l (MAFF, 1991). For England and Wales it is estimated that farm livestock produce 2.5 million tonnes of BOD every year. If just 2% enters inland waters it would be equivalent to the total BOD loading of treated effluents from sewage treatment works (NRA, 1992). Considerable research has also been undertaken on the diffuse input of nitrates and phosphates to water courses from agricultural land largely as a result of the impact of high nutrient loadings on receiving water courses (Addiscott *et al.*, 1991; Burt *et al.*, 1993; NRA, 1992).

Whilst the polluting potential of liquid slurries, nutrients and pesticides has been well researched over the last decade, little attention has focused on other wastes produced by a range of farming activities. Agriculture uses a diverse array of inputs and produces a plethora of lesser wastes. Amongst these are spent sheep dip, pesticide residues, of which there are over 500 active ingredients in use (BCPC, 1995), containers, packaging, animal carcasses, veterinary products, string, drums, waste oil and used tyres. It is only comparatively recently that the potential for pollution to air, water and soil from inappropriate disposal of these wastes has been recognised (Lowe and Ward, 1993). Apart from the high BOD and undesirable nutrient content of farm wastes, many are extremely toxic and can do severe damage to riparian ecosystems (NRA, 1992). There is an additional threat of pathogens from carcasses and other animal materials infecting water supplies following inadequate disposal.

Farm pollution represents around 12% of substantiated water pollution incidents reported annually (NRA, 1993). The majority of farm waste pollution incidents are attributed to organic wastes (87%), oil (3%), pesticides (< 2%), nutrients (< 1%) and others (ca. 7%) (NRA, 1992). In the same report, and from over 1000 farm inspections, the NRA suggests that the proportion of farms polluting, or at high risk of doing so, is about 40% of the total. Undoubtedly many pollution incidents are unreported.

The widespread use of soakaways to dispose of sheep dip and pesticides was a recommended method until recently. As a consequence, organic pollution of ground-waters is thought to be widespread in dairy farming areas (Seger *et al.*, 1992) and pesticide and dips are widely detectable in drinking water supply zones (ENDS, 1993a, b). Air and soil can be polluted by uncontrolled burning and/or burial of waste, although this has received much less attention (Lowe *et al.*, 1990). Prosecutions for pollution of air and soil are much less common than for water.

At the beginning of the 1990s, the UK Government set itself a range of environmental targets in *This Common Inheritance: Britain's Environmental Strategy* (HMSO, 1990), including the integration of environmental considerations with economic activity and adoption of the "Polluter Pays" principle. The emphasis on waste rests on the ethos of reducing the use, reusing or recycling of materials by ensuring stricter controls over waste disposal.

Despite the recent movement towards a "post-productivist" ethos in EU agriculture, usually conceptualised as the post-productivist transition (Shucksmith, 1993; Lowe *et al.*, 1993; Ilbery and Bowler, 1998), successive governments have encouraged and supported agricultural modernisation through systems of guaranteed prices, subsidies and grants. One of the consequences has been a major restructuring of agricultural production in which intensive farming has become concentrated on larger, more specialised units and in fewer regions in the EU (Bowler, 1985). Even as late as 1980, the Common Agricultural Policy (CAP) was advocating further structural change, and increased productivity was perceived to be the means of securing growth in output (HMSO, 1980). Burgeoning costs and more general environmental motives are now beginning to influence policy to a greater extent, and agricultural support has increasingly become conditional on implementing agri-environmental considerations (HMSO, 1992; Wilson, 1996; Potter, 1998). Their growing importance is a striking demonstration of the shift in parliamentary and public attitudes towards agriculture in the EU (Rolfe, 1990). However, these considerations have been widely criticised (Potter, 1993; Wynne, 1992) and tend to relate only to the conservation of countryside features and wildlife, which are highly visible elements of the environment. Less emphasis has been placed on the more unobtrusive aspects of agriculture, such as waste, and even measures to reduce production can have adverse environmental effects. For example, the introduction of milk quotas increased production intensity, in turn leading to further pollution pressure (Halliday, 1988; Lowe *et al.*, 1992). Indeed, in a survey of dairy farms, pollution regulation was perceived to be the major problem (Centre for Agri-Food Business Studies, 1991).

Lowe *et al.* (1992) and Clark *et al.* (1994) have criticised farm waste regulations and pollution control policies aimed at dairy farms, and have expressed doubts that they will either be enforced or reduce pollution. Furthermore, Lowe *et al.* (1992) conclude that, while dairy production is becoming increasingly unacceptable to the National Rivers Authority (now the Environment Agency), farmers treat it purely as a technical problem and the Agricultural Development and Advisory Service (ADAS) as a matter of fact. Ward and Lowe (1994) suggest that farmer succession creates a greater likelihood of investment in pollution prevention equipment, although they admit that this is not an accurate indication of the soundness of farmers' environmental practices. Ward *et al.* (1995) also suggest that farmers' attitudes to pollution and regulation vary greatly, whilst Barker (1991) found farmers to be resentful, unco-operative and with a sense of apathy to pollution. Recent studies of farmers' attitudes towards environmental protection policy through the designation of Sensitive Areas (Foster *et al.*, 1989; Foster and Ilbery, 1992; Wilson, 1996) emphasise the gulf between best practice and advice provided by government agencies and the situation on the ground. Lack of awareness of regulations, and the necessity to increase production, have outweighed environmental concerns. These studies conclude that the relationship between policy makers and regulators on the one hand and the farming community on the other hand is usually weak and often unrecognisable.

Individual farmers vary enormously with regard to the extent to which they search for, and use, information (Foster *et al*, 1989; Fearne, 1990). Interpersonal contact between farmers remains as one of the most important channels of information (Jones, 1975) and it has been found that the most important source of influence on farmers' conservation practices is from within the farming community itself, or closely associated with it (Carr and Tait, 1991). Farmers tend to gain their initial awareness of a problem from the mass media (Read, 1987), although commercial representatives and agents are becoming increasingly important as sources of advice and information. It has also been suggested that older farmers pay for advice only as a last resort, tending to rely on passive information, whilst the younger, better educated farmers, and those on larger farms, are more willing to seek, use and pay for information (Fearne, 1990). Farmer behaviour still tends to be dominated by economic and farming considerations rather than the environment. Many decisions, especially those of a day-to-day nature, are habitual and based on solutions that worked in the past (Carr, 1988).

It is evident from this brief review that the farming community is a highly studied social group. However, little is known about non-organic waste disposal practices currently used by farmers. The following sections summarise the results of a recent (1994) pilot questionnaire survey on waste-disposal perceptions, attitudes and practices on a sample of 48 farms in the Stratford district of south Warwickshire. This survey considered four major issues associated with waste amongst the farming community in this district:

- waste awareness and understanding;
- waste disposal practices;
- legislative awareness and compliance; and
- information and advice on farm waste.

THE FARM SURVEY

A random sample of 48 farms was selected from a list of farmers drawn from the Yellow Pages Directories for the Stratford district of south Warwickshire (Figure 13.1). Although there is some discussion concerning the efficacy of Yellow Pages as a sampling frame (Errington, 1985; Emerson and MacFarlane, 1995), comparison of the selected farms with the most recently available MAFF census data at the individual parish level suggests that the sample is a reasonable representation of the distribution of cropping practices throughout the district for farms greater than 10 ha in area (Table 13.1). However, it is acknowledged that changes in farm enterprise (especially the increase in rape production at the expense of other break crops and the introduction of set-aside) occurred between 1988 and the time of the survey. Detailed parish-based statistics are available from MAFF only up until 1988 (Evans, 1996) and other sources of information at this level of detail are not in the public domain. Animal numbers compare less well with the MAFF census (Table 13.1) and pig and sheep farms are slightly under-represented in the sample. The number of sample farms by dominant enterprise is given in Table 13.2.

A detailed questionnaire containing 72 specific questions was designed to obtain information on the four major issues identified above. Closed and open questions were included, the latter to allow farmers to express concerns in their own way. The interviews were conducted face-to-face with the 48 farmers.

FARM WASTE: SURVEY FINDINGS

Results presented below deal with the four main issues identified in the introduction, namely waste awareness and understanding; waste disposal practice; awareness of, and

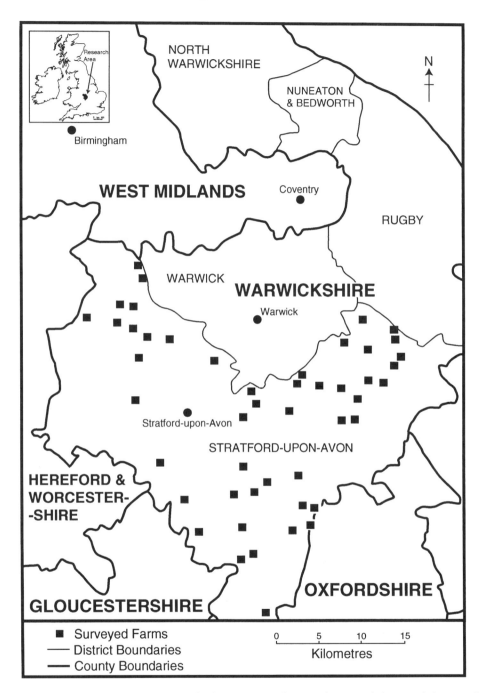

Figure 13.1 *Location of the Stratford-upon-Avon district of Warwickshire and the sampled farms*

Table 13.1   *Agricultural land use and livestock comparisons between the MAFF (1988) census and the farm sample (1994)*

Land use	Cropping area (%)	
	MAFF 1988 (Stratford district)	1994 survey ($n = 48$)
Grass	41.5	40.8
Cereals	41.8	41.4
Roots	0.8	0.7
Horticulture	1.3	0.0
Pulses	3.6	3.5
Rape	4.9	6.8
Fruit/vegetables	1.3	0.0
Set-aside	—	6.8
Fallow/misc.	4.8	—

Livestock	Number in district	Number in survey	Sample % of MAFF totals
Dairy	19 380	1941	10.0
Beef	24 629	2189	8.9
Pigs	24 262	13	0.05
Sheep	204 109	8160	4.0
Chickens	671 424	57 000	8.5

Table 13.2   *The number of sample farms by dominant enterprise*

Sheep/cattle	13
Dairy	14
Mixed	9
Cereals/crops	14
Poultry	1
Total	48

attitudes towards, waste legislation; and sources of information and advice regarding farm waste.

Waste Awareness and Understanding

Of the total sample, only two farmers identified environmental concerns as an important general agricultural issue and only one identified inorganic farm wastes (particularly plastics) as a problem. When focusing on the problem of farm waste, greatest concerns were expressed by dairy farmers (72.7%) which is in contrast to the responses from sheep/beef cattle producers (7.7%). Silage producers were almost eight times more likely than other groups to identify pollution as a potential problem. However, pollution is not perceived to be a major issue. When asked to rank pollution on a scale of 1 to 6 in comparison with other agricultural concerns, the average score of 3.1 demonstrates a lack of significance placed on pollution issues. Furthermore, and based on the same scaling system, the polluting potential of farm wastes was scored 3.5, suggesting that farmers were often not aware of the serious polluting potential of wastes like milk and sheep dip.

Organic wastes were considered by many farmers to be an asset, whilst older farmers, and those farming small units, found most difficulty in identifying waste products. Part-time farmers also showed a poor awareness of wastes, with five out of seven not identifying any specific waste product despite the fact that all farmed sheep and had sheep dip waste. The wastes named by farmers are listed in Table 13.3 where it is clear that plastics give the greatest concern.

Waste Disposal Practice

Farmers disposed of inorganic wastes in a variety of ways. Of the 11 types of inorganic waste identified, many disposal methods were generally unsuitable if not illegal (Table 13.4). Burning was one of the most popular disposal methods for a range of wastes including tyres and plastics, despite specific codes of practice provided by MAFF (1992) recommending that burning is not a practicable option. Paragraph 145 (MAFF, 1992) states: "do not burn plastics in the open as this can cause large amounts of dark smoke and poisonous by-products". Only four farmers recycled plastics, despite the fact that almost 50% of the respondents were aware of the existence of a recycling scheme.

Another popular, but dubious, option was the dumping or disposal of undiluted wastes (especially sheep dip) and veterinary products in pits or ditches; this again is specifically in breach of MAFF (1991) and veterinary (Veterinary Medicines Director-ate, 1991) codes of practice. MAFF (1991) paragraph 210 states that soakaways are "not suitable in most places". Two farmers dumped milk, one into a drain and the other into a pond. The latter commented that this was not a problem "as long as there were ducks on his water".

Some innovative methods had been developed for the disposal of sheep dips. One farmer used the dip as a straw steriliser, whilst 40% of those using dips simply pulled

Table 13.3 *Wastes identified by the Strat-ford district farmers (as a percentage of sample interviewed)*

Plastics	68.3
Muck/slurry	46.3
Spray cans	41.5
Effluents	26.5
Spray washings/sheep dip	24.4

Table 13.4   Disposal methods matrix for farm wastes

Disposal method	Waste type*										
	A	B	C	D	E	F	G	H	I	J	K
None produced	6	2	15	23	12	11	25	35	3	1	5
With household	26					2	5				
Take-back	1								19		
Recycle	1	1	3				4			9	
Reused/sold	2	18	1	1		6		5	17	23	
Stored	2					1	4		3	2	
Land spread				14	26			6			
Buried	3	1	1			3	8				6
Burned	5	26	28			24			9	11	3
Dumped/pit	2			10	10	1	2	2		1	
Kennels/knacker											27
Kennels/burned											6

* A, medicinal/veterinary; B, fertiliser bags; C, silage wrap/sheets; D, sheep dip; E, sprayer washings; F, plastic chemical cans; G, metal chemical cans; H, milk; I, tyres; J, oil; K, dead stock
Number of replies: 4
Number of replies with unsuitable/illegal method: **4**

the plug on the dip trough. Whilst many farmers were aware of the change from organophosphorus to non-organophosphorus dips, it was commonly assumed that the latter products were less potent and that disposal by the methods outlined above was not unreasonable. Despite the fact that non-organophosphorus dips are less persistent, Austin (1995) argues that their disposal in concentrated form can still cause immense ecological damage.

It is evident from the questionnaire analysis that the disposal methods given in Table 13.4 were adopted either because of the ease or practicability of the method or because it was the cheapest. Of the 391 responses where waste is produced, 173 (44%) posed a potential threat to the environment as being either unsuitable or illegal. Furthermore, it was the older (> 40 years of age) farmers who persistently disposed of inorganic wastes using methods which posed greatest environmental risk.

Awareness of, and Attitudes Towards, Waste Legislation

In addition to the existence of various codes of good agricultural practice produced by MAFF and related organisations, other government and non-government schemes exist for the disposal of wastes and the promotion of health and safety. These include plastic recycling and the national pesticide retrieval scheme. However, out of the sampled group only 6% had recycled plastics and 14.6% had used the pesticide retrieval scheme. Just five farmers had completed Farm Waste Management forms and eight had done a COSHH (Control of Substances Hazardous to Health) assessment;

one of these had copied his neighbour's return! Furthermore, although veterinary and chemical treatments and stocks should be recorded by law, almost one-third who either sprayed or had stock kept no records.

Whilst the majority had heard of one of the three MAFF codes of good agricultural practice, only seven farmers owned a copy of one or more of these codes. The general response was that they "would not read them as there were better things to do". They were perceived to contain good advice but, in general, were "too long-winded". Over 60% of the respondents honestly accepted that they had knowingly broken codes regarding waste disposal. Burning dominated the known "transgressions", with chemical dumping accounting for more than 15%. The primary reasons given for breaching codes were that there was no alternative and that it was a function of the ease and practicability of the current method used.

The overwhelming conclusion from an analysis of the responses is that existing advice, information and/or legislation is largely being ignored by the agricultural community. It was argued by the majority of the respondents that better or alternative methods of disposal would reduce code violations. Getting caught came a close second, highlighting the fact that most farmers were well aware that existing codes were being broken.

Sources of Information and Advice Regarding Farm Waste

Whilst many sources of information and advice were cited by respondents, their first choice for information was trade journals followed by word of mouth and advice from MAFF (Table 13.5). Representatives and consultants were barely mentioned. Most of the sample believed that farm waste information and advice were channelled to farmers only moderately well. This varied according to farm type, with cereal and dairy farmers being more impressed than sheep and cattle farmers.

There was considerable reluctance amongst those interviewed to seek advice directly from either ADAS or the NRA (now the Environment Agency) mainly for fear of follow-up visits. Farmers approaching the agencies for advice were generally positive. The youngest, and mainly dairy farmers, were the most likely to use ADAS, while arable farmers in the 50 to 60 age range preferred the National Farmers' Union (NFU). Those who had little contact with the agencies were generally

Table 13.5 *Sources of information used by farmers (number of responses)*

Source	1st coice	2nd choice	3rd choice
Trade journals	30	9	5
Word of mouth	5	10	5
Circulars/mailshots	1	9	5
NFU	3	5	4
ADAS	4	3	4
MAFF	5	5	1
Consultants/reps	—	2	1
NRA (EA)	—	2	—

unwilling to approach them and had negative attitudes towards the possibility of obtaining advice.

DISCUSSION AND CONCLUSION

The general lack of interest in the subject of farm waste shown by the sampled farmers, coupled with a blatant disregard for existing legislation and codes of good agricultural practice, raises a number of significant issues relating to the impact of such practices on surface water and groundwater quality and the potential for both soil and air pollution. Most farmers cared little about the problem of waste management and disposal, supporting the findings of Barker (1991) and Ward *et al.* (1995). The research also supports the general conclusions reached by Carr (1988) that farmer behaviour is still dominated by economic and farming considerations; day-to-day decisions are habitual and given little thought.

It is evident that few farmers are utilising existing schemes for plastics recycling and chemical take-back. The burning, dumping and disposal of inorganic wastes, sheep dip and veterinary products in ditches, lagoons and drains in an undiluted form are common methods of waste disposal. Farmers do not perceive pollution as a problem and generally do not appreciate the polluting potential of farm wastes. The disposal methods are employed because they are practical and cheap. Most farmers disposing of wastes do so illegally in the knowledge that they are in breach of codes of good agricultural practice. There is a reluctance amongst the farming community to approach the regulatory authorities for advice and guidance. This reluctance stems in part from fear of prosecution. Glasbergen (1992) and Lowe *et al.* (1992) argue that agricultural waste policy is a containment and education exercise which does not eradicate the waste problem. The Silage, Slurry and Agricultural Fuel Oil Regulations, for example, specify criteria for storing wastes rather than reducing them. This may have allowed some complacency to develop in those farmers who have received an inspection visit or who had upgraded facilities. Few seemed to realise that good waste administration depends on daily management.

The overall conclusions derived from this investigation support many of those presented in other studies of farmers' attitudes towards pollution regulation. Despite the availability of advice, codes of practice and legislation, farmers are either not aware of, or choose to ignore, such information. The greatest threat to the environment, therefore, appears to derive from the inadequate link between advice on best practice and the waste disposal options currently used by the farming community.

The results of this pilot project have broader implications for the control of agricultural pollution. Current methods employed to regulate pollution from agriculture are inadequate and new innovative methods are required to improve environmental awareness in the agricultural community. These might lead to a personal desire to minimise the environmental threats posed by farm waste disposal.

ACKNOWLEDGEMENTS

The south Warwickshire questionnaire survey was undertaken for the production of an MSc thesis by Mark Hancock entitled *Farm Waste Management and Policy: an Initial Study of South Warwickshire Farmers*. It was completed in partial fulfilment of the Coventry University MSc programme "Environmental Monitoring and Assessment". We would like to express our sincere thanks to those farmers who participated in the survey, the detailed results of which will remain confidential.

REFERENCES

Addiscott, T.M., Whitmore, A.P. and Powlson, D.S. 1991. *Farming, Fertilisers and the Nitrate Problem.* CAB International, Wallingford.

Austin, R. 1995. Pollution is far worse with non-OP dip disposal. *Farmers Weekly*, (18 August), 37.

Barker, P. 1991 Agricultural pollution: control and abatement in the upper Thames region. *Journal of the Institute of Water and Environmental Managers*, **5**, 318–325.

BCPC (British Crop Protection Council). 1995. *The UK Pesticide Guide*. CAB International, Wallingford.

Bowers, J. 1995. Sustainability and agricultural policy. *Environment and Planning A*, **27**, 1231–1245.

Bowler, I.R. 1985. Some consequences of the industrialisation of agriculture in the European Community. In Healey, M. and Ilbery, B. (Eds), *The Industrialisation of the Countryside*. GeoBooks, Norwich, 75–98.

Burt, T.P., Heathwaite, A.L. and Trudgill, S.T. (Eds), 1993. *Nitrate: Processes, Patterns and Management*. Wiley, Chichester.

Carr, S. 1988. *Conservation on farms: conflicting attitudes, social pressures and behaviour*. Unpublished PhD Thesis, University of Oxford.

Carr, S. and Tait, J. 1991. Farmers' attitudes to conservation. *Built Environment*, **6**, 218–223.

Centre For Agri-Food Business Studies. 1991. *A Survey of Dairy Farmers: attitudes towards MMB reforms*. Royal Agricultural College, Cirencester.

Clark, J., Lowe, P., Seymour, S. and Ward, N. 1994. *Sustainable agriculture and pollution regulation in the UK*. Working Paper Series WP13, Centre for Rural Economy, University of Newcastle upon Tyne.

Cox, G., Lowe, P. and Winter, M. 1986. Agriculture and conservation in Britain. In Cox, G., Lowe, P. and Winter, M. (Eds), *Agriculture: People and Policies*. Allen and Unwin, London, 181–216.

Emerson, H. and MacFarlane, R. 1995. Comparative bias between sampling frames for farm surveys. *Journal of Agricultural Economics*, **46**, 241–251.

ENDS. 1993a. *Research underlines risks from sheep dip chemicals*. March, 8–9.

ENDS. 1993b. *Sheep dip problems*. August, 32.

Errington, A. 1985. Sampling frames for farm surveys in the UK: some alternatives. *Journal of Agricultural Economics*, **36**, 251–258.

Evans, N. 1996. Evaluating recent changes to the publication of UK agricultural census data. *Geography*, **81**, 225–234.

Fearne, A.P. 1990. Communications in agriculture: results of a farmer survey. *Journal of Agricultural Economics*, **41**, 371–381.

Foster, I.D.L and Ilbery, B.W. 1992. Water protection zones: a valid management strategy? In Gilg, A.W. (Ed.), *Restructuring the Countryside: Environmental Policy in Practice*. Avebury, Aldershot, 203–220.

Foster, I.D.L., Ilbery, B.W. and Hinton, M.A. 1989. Agriculture and water quality: a preliminary investigation of the Jersey nitrate problem. *Applied Geography*, **9**, 95–114.

Glasbergen, P. 1992. Agro-environmental pollution: trapped in an iron law. A comparative analysis of agricultural pollution controls in the Netherlands, United Kingdom and France. *Sociologia Ruralis*, **32**, 30–48.

Halliday, J. 1988. Dairy farmers take stock, a study of milk producers' reaction to quota in Devon. *Journal of Rural Studies*, **4**, 193–302.

Hawkins, K. 1984. *Environment and Enforcement: Regulation and the Social Definition of Pollution*. Clarendon Press, Oxford.

House of Commons Environment Committee. 1989. *Toxic Waste*. Second Report 1988/89, HMSO, London.

HMSO. 1980. UK Government 2nd White Paper *Farming and the Nation*. HMSO, London.

HMSO. 1990. *This Common Inheritance: Britain's Environmental Strategy*. HMSO, London.

HMSO. 1992. Agri-Environmental Regulations, June 1992. *Official Journal of the EEC* L215/85, 30/7/92. HMSO, London.

Ilbery, B.W. and Bowler, I.R. 1998. From agricultural productivism to post-productivism. In Ilbery, B.W. (Ed.) *The Geography of Rural Change*. Addison Wesley Longman, London, 57–84.

Jones, G.E. 1975. *Innovation and Farmer Decision Making. Module D203: Agriculture*. Open University Press, Milton Keynes.

Lowe, P. and Ward, N. 1993. *Risk, Morality and the Social Construction of Farm Pollution, Engaging the Risk Society*. Paper presented to the *Annual Conference of the Rural Economy and Society Study Group*, Royal Holloway and Bedford New College, London. December 1993.

Lowe, P., Cox, G., Goodman, D., Munton, R. and Winter, M. 1990. Technological change, farm management and pollution regulation: the example in Britain. In Lowe, P., Marsden, T. and Whatmore, S. (Eds), *Technological Change and the Rural Environment*. David Fulton, London, 53–80.

Lowe, P., Clark, J., Seymour, S. and Ward, N. 1992. *Pollution Control on Dairy Farms: an evaluation of current policy and practice*. SAFE Alliance, London.

Lowe, P., Murdoch, J., Munton, R. and Flynn, A. 1993. Regulating the new rural spaces: the uneven development of land. *Journal of Rural Studies*, **9**, 205–222.

MAFF. 1991. *Code of Good Practice for the Protection of Water*. Ministry of Agriculture, Fisheries and Food, London.

MAFF. 1992. *Code of Good Practice for the Protection of Air*. Ministry of Agriculture, Fisheries and Food, London.

Munton, R., Marsden, T. and Whatmore, S. 1990. Technological change in a period of agricultural adjustment. In Lowe, P., Marsden, T. and Whatmore S. (Eds), *Technological Change and the Rural Environment*. David Fulton, London, 104–126.

NRA. 1992. *The influence of agriculture on the quality of natural waters in England and Wales*. Water Quality Series No. 16, National Rivers Authority, Bristol.

NRA. 1993. *Water pollution incidents in England and Wales 1992*. National Rivers Authority, Bristol.

Potter, C. 1993. Pieces in a Jigsaw: a critique of the new agri-environment measures. *Ecos*, **14**, 52–54.

Potter, C. 1998. Conserving nature: agri-environmental policy development and change. In Ilbery, B.W. (Ed.), *The Geography of Rural Change*. Addison Wesley Longman, London, 85–105.

RCEP (Royal Commission on Environmental Pollution). 1985. *Managing Wastes: the Duty of Care*. HMSO, London.

Read, N. 1987. Changing sources of information on farming. *Farm Management*, **6**, 289–296.

Rolfe, N. 1990. British Agricultural Policy and the EEC. In Britton, D. (Ed.), *Agriculture in Britain: Changing Pressures and Policies*. CAB International, Wallingford, Ch. 10.

Seger, J., Jones, F. and Rutt, G. 1992. Assessment and control of farm pollution *Journal of the Institute of Water and Environmental Management*, **6**(1), 48–54.

Shucksmith, M. 1993. Farm household behaviour and the transition to post-productivism *Journal of Agricultural Economics*, **44**, 466–478.

Veterinary Medicines Directorate. 1991. *The safe handling and disposal of sheep dips.* Advisory note to farmers. MAFF, London.

Ward, N. and Lowe, P. 1994. Shifting values in agriculture: the farm family and pollution regulation. *Journal of Rural Studies*, **10**, 173–184.

Ward, N., Lowe, P., Seymour, S. and Clark, J. 1995. Rural restructuring and the regulation of farm pollution. *Environment and Planning A*, **27**, 1193–1213.

Waste Management Licensing Regulations. 1994. SI 1994/1056. HMSO, London.

Wilson, G. 1996. Farmer environmental attitudes and ESA participation. *Geoforum*, **27**, 115–131.

Wynne, P. 1992. The missed opportunities of CAP reform. *Ecos*, **13**, 20–24.

14

Screw the Lid even Tighter? Water Pollution and the Enforcement of Environmental Regulation in Less Developed Countries

Stuart N. Lane, Keith S. Richards, Sudanshu Sinha and Shuang-ye Wu

INTRODUCTION

Baker (1990) suggests two reasons for the failure of environmental policy: (a) regulatory, in which there is a general absence of a legal mandate for environmental protection; and (b) enforcement, in which administrative infrastructure is largely non-existent or inadequate, leading to "administrative traps" and ultimately policy failure. At all scales of concern over environmental protection, there is emphasis upon development of environmental policy, but issues of enforcement, and in particular its feasibility and practicality, have been given less attention. With respect to the water environment, in More Developed Countries (MDCs), once legislation is developed it is generally reasonably well enforced, facilitated both by effective regulatory agencies and by a culture that is adapted to the idea that certain behaviour should be defined as deviant and regulated to allow better environmental protection. This attitude is facilitated by the long time period over which legislation has been developed and the fact that MDCs can afford to divert scarce economic resources into environmental protection. However, the situation in Less Developed Countries (LDCs) is somewhat different. Pollution regulation is a relatively new concern, and is being introduced more rapidly and comprehensively, at least in terms of legislation, than it was in

Water Quality: Processes and Policy. Edited by Stephen T. Trudgill, Des E. Walling and Bruce W. Webb.
© 1999 John Wiley & Sons Ltd.

MDCs. There is much faith placed in the development of appropriate legislation for pollution control, often mirroring legislation developed in MDCs and associated with both a colonial legacy and a continual exposure of key legislators in LDCs to MDC education, academics and policy makers, and hence legislative principles and procedures. This has resulted in some extremely sophisticated anti-pollution regulation. However, there remain continual examples of serious violations of pollution regulations (e.g. Chakraborty, 1995; Sharma, 1996; Banerji, 1996; Banerji and Martin, 1997) that observers argue are the norm rather than exceptions to the rule (e.g. Banerji and Martin, 1997).

Continual violation of this sort implies that attention must be given to the process of law enforcement in specific contexts. This paper aims to evaluate the enforcement of water pollution policy in India to begin to understand what developments are required to improve water quality. The central tenet is that improved enforcement does not simply require a stricter application of existing pollution regulations, as this would simply reinforce weaknesses already present in reliance upon regulations *per se* as the means of ensuring adequate environmental protection. Rather, there must be continued development of effective administrations to allow both alternative enforcement methods and the development of mutual trust between regulator and regulated. Such trust begins to allow more compliance-based approaches to develop. The paper begins by introducing some basic legal principles, and the specific water pollution legislation introduced in India. It then reviews existing approaches to pollution control in India, which tend to rely heavily upon sanctions and their enforcement. It considers the possibility of a more compliance-based approach, but then argues that without more effective administrative systems, such an approach is unlikely to work.

PRINCIPLES OF REGULATION

Issues surrounding basic principles of regulation need to consider: (i) the intention of regulation; (ii) the associated nature of the regulations required; and (iii) the means by which they are enforced. In practice, these three issues are not independent of one another. Regulatory intention can be broadly divided into economic and ethical views (e.g. Sagoff, 1988). Following Hawkins (1984), the essence of water pollution control is the development of pollution regulations and laws that either: (i) regulate economic and social activities in an attempt to minimise their negative impact on the water environment; or (ii) directly exclude certain types of activity whose effects are regarded as sufficiently harmful or undesirable to warrant complete exclusion. Implicit in both of these is the extension of public authority over organisational and business life on the assumption that pollution represents a specific form of market failure, which requires direct intervention as a consequence. The regulation process is largely based upon a theoretical, and classical, economic perspective: the "Rational Polluter Hypothesis". Given the neoclassical economic theory of profit maximisation, it is thought that a polluting industry weighs the financial costs and benefits of environmental compliance and decides to act only when benefits dominate (Bradbaart, 1995). The role of legislation is, therefore, to determine the maximum acceptable level of pollution in the system under consideration, and to choose a level of cost that is sufficient to reduce polluting activities to that level. However, in addition to this largely economic view of legislation, in which environmental problems are seen as little more than negative externalities, there is also an important ethical view (e.g. Sagoff, 1988). Sagoff argues that many environmental laws balance the relative merits of environmental protection, in a conventional cost–benefit analysis framework. However, consequent debate over appropriate levels of protection, combined with the methods that have to be invoked to allow intangibles (e.g. aesthetic values) to be incorporated into an economic analysis, mean that environmental laws embody a much broader rationality and ultimately purpose. Sagoff (1988, p. 97) notes that environmental "...laws demonstrate that we are not consumers simply bent on satisfying every subjective preference..." but also citizens adopting "...a model of government and a vision of political life that allow us to posit collective values and to give effect to our national conscience and common will...". Environmental pollution becomes an evil which must be eliminated to publically defined acceptable levels to allow society to live up to its ideals and expectations (Sagoff, 1988, p. 196).

The distinction between economic and ethical support for environmental law is important, as each implies different intentions for regulatory development, and may therefore result in different types of regulation and different enforcement strategies. From an economic perspective, legislation is technology-forcing, resulting in regulations that aim to eliminate environmental damage by making polluting activities expensive, but not necessarily a criminal act, even if subsequent legislation makes it a criminal offence to exceed established levels. From an ethical viewpoint, there is a much stronger sense that pollution is defined as an unacceptable harm committed against society as a result of the consequences it has for both present and future (e.g. Sagoff, 1988), with regulations that should be specified in criminal terms.

Regulatory development and enforcement are closely connected and may be divided for convenience into sanctions-based and compliance-based, although such a division is artificial as both are end members of what is really a continuum, and any enforcement approach may contain elements of both. Sanctions-based approaches aim to enforce regulation by compulsion and coercion, with a penal approach to dealing with deviant activity. Black (1976) describes a penal approach as one that emphasises the prohibition of certain conduct, in which prohibition is enforced by punishment. Regulations exist to turn previously legitimate conduct into deviant activity (Hawkins, 1984). Enforcement is reflective, based upon identification of deviant activity, determination of what harm has been done, detection of the law-breaker, and choice of an appropriate solution to the problem. The identification, determination and detection stages in a sanctions-based system are linked with a special concern for proof of violation (Reiss and Biderman, 1980) through adjudication with outcomes that are binary in nature (guilty or not-guilty). This is a complex legal process (e.g. Padfield, 1995) involving the state attempting to prosecute a defendant, and necessarily relies upon proof of strict or absolute liability under criminal law. The victim of a pollution event is not involved in the legal process, except in providing evidence where appropriate, although they may be eligible for compensation as a result of a conviction. This has the advantage for the victim of no risk: if the court were to rule in favour of the defendant, the victim would incur no costs. However, it has the disadvantage of tending to reduce the amount of compensation the victim receives, as the fines imposed on offenders tend to be limited under absolute or strict liability, partly because of the importance of precedent in determining the magnitude of compensation received. In the case of water pollution, environmental restoration (if it is possible) tends to require technology and is therefore expensive. Thus, it is clearly in the interests of the victim to obtain maximum compensation, to assist with both the negative consequences of exposure to the pollution (e.g. ill-health) and to aid with any remediative measures.

The alternative to prosecution by the state is personal litigation by the victim under civil law, wherein a plaintive sues the defendant responsible for causing the damage experienced by the plaintive. This involves the principles of tort, in which damages are awarded that may not only be punitive (as is normally the case under criminal law) but also exemplary. Thus, the victim may receive larger amounts of remuneration from the defendant, in proportion to the magnitude of damage determined by the Court, but runs the risk of incurring significant cost if the Court rules in favour of the defendant. It is an important development from a sanctions-based approach, as the emphasis switches from proving deviance to proving damage. Damage is defined independently of any sanctions established as a part of the regulatory process, although the existence of sanctions may be central to determining guilt prior to determining the magnitude of damage.

Once regulations are in place, it is assumed that polluters will be deterred from engaging in deviant activity by the threat of criminal proceedings if they continue to do so. Underpinning a sanctions-based approach is a techno-centric attitude which assumes that pollution can be readily defined, viewed as largely a technical problem, and tackled by "modern" technology, both in the enforcement of standards and the identification of deviance. The approach is also broadly "modernist" in terms of its legal principles and necessary administrative structures: even if the legislation does not involve technology directly, it follows the modernist dictate that elements of the social world can be directly or indirectly manipulated or controlled to achieve a given end in the natural world. There is no dispute as to what that end should be or how it should be achieved. According to O'Riordan (1981), this approach is based upon rationality, managerial efficiency and a sense of optimism or faith in human ability to understand and control nature. This approach has been the subject of numerous behavioural studies into its effects on environmental decision making (Caldwell, 1970, 1987; Mitchell, 1971; Sewell, 1971) with the conclusion that it is a mutually self-reinforcing paradigm which excludes critical evaluation of competing and technology-undermining schools of thought. It results in knowledge systems that support a "tyranny of expertise" (Hart, 1988), disregarding the opinions of the public and local knowledge of local environments. Indeed, sanctions-based approaches may be reinforced by the separation of decision makers from the localities that they attempt to regulate, such that there may be little appreciation of the need for, and difficulties of, pollution regulation in particular local contexts.

Under a sanctions-based approach, the existence of regulation is assumed to be the only prerequisite for effective improvements in environmental quality, ignoring or only partially recognising the critical importance of compliance as part of the regulatory process. Indeed, compliance without punishment may be unacceptable from an ethical view point as if the activity has been defined as punishable there is a moral duty to ensure that deviant activity is punished. Compliance-based approaches differ from sanctions-based approaches as they begin by recognising that the detection of deviant activity is the first step towards enhancing prevention, rather than prosecuting deviance. Punishment is thought to be an unsatisfactory operational philosophy as it removes control of the case from those involved in improving environmental quality and does not necessarily address the causes of deviant activity. In a more compliance-based approach, regulatory agencies take given environmental goals and determine the best ways of meeting them. Pollution control officers are then responsible for negotiating with polluters in an attempt to reduce or eliminate pollution discharge, and so attain the socio-economic goal of improvement in environmental quality at least economic cost to the polluter. Different polluters will be more or less able to engage in pollution control, and there will be differing needs for them to do so, according to the assimilative capacity of the environment into which they discharge. Thus, the emphasis becomes local and emancipatory, involving education of both polluters and polluted to the benefits of pollution control, in an attempt to make continued polluting activity unattractive. Regulations may exist as part of this process (to define minimum acceptable standards for emissions or environmental quality), and may be the ultimate recourse for pollution control officers. Thus, decisions in a compliance-based system are graded in charter (Eisenberg, 1976), and although in rare cases matters are ultimately settled by adjudication, they are normally controlled by the parties themselves in private negotiations which rely on bargaining, not adjudication. Adjudication, and the associated legal process, is the end-point that a pollution control authority would make recourse to when all other approaches fail. Compliance-based approaches have as their prime emphasis the improvement of environmental quality rather than ensuring punishment of those creating environmental damage. However, they implicitly accept that environmental damage is a different sort of crime which may be ethically unacceptable. There remains the continual concern that co-operation may become collusion, and that the enforcement style will then fail to result in improvements in environmental quality or reductions of environmental damage.

Thus, whether or not one holds an economic or ethical view for the role of water pollution regulation, what really matters is the view held about enforcement, and the ability of different enforcement procedures to exist within a given society.

SHAPING THE LID: WATER POLLUTION REGULATION IN INDIA

India was the first country to change its constitution to protect its environment (Dwivedi and Khator, 1995). Since then, the development of laws to protect the environment has been rapid, reflecting the basic belief that regulation and enforcement is the best means of achieving implementation, with a legal foundation that is largely a hybrid of US and UK legal systems (Dwivedi and Khator, 1995). The functions of this legislation are: (i) to extend existing penal code, which itself makes fouling water used for drinking purposes a criminal offence, to all cases in which water bodies receive polluted effluents; (ii) to require consent for all discharges to the water environment; (iii) to allow for development of an institutional structure, including its financing, for implementation of pollution regulations; (iv) to give such institutions relevant powers, including standard setting and bringing forward prosecution; and (v) to deal explicitly with interpretations of criminal justice such as absolute liability in the specific context of polluter regulation. This last issue is necessary to define responsibility for pollution where definition of a guilty party (an employee, an employer, or an institution) has to move beyond the traditional jurisdiction of criminal justice.

The development of pollution legislation in India has focused upon defining pollution as a criminal act. The most important legislation was the Water (Prevention and Control of Pollution) Act (1974), which first established nationwide integrated pollution control, even though its incorporation was achieved on a state-by-state basis. This established Pollution Control Boards (Central and State) as unitary agencies for the prevention and control of water pollution and created Central and State water testing laboratories to assess the extent of pollution, to lay down standards and to establish guilt or default of provisions of the Act. This Act also gave the State Boards the right to monitor any water body, including effluent-bearing bodies. In the case of dispute over the quality of an effluent-bearing body, it gave provision for one sample to be analysed by a laboratory recognised by a State board, and a second to be analysed by a recognised laboratory requested by the land owner. The Boards' powers included entry and inspection for persons appointed by the Board. The Act prohibited persons from knowingly causing or permitting the disposal of polluting matter, as defined by standards laid down by the State Board, into any stream or well, and prohibited the creation of new industries which would discharge effluent into water bodies, as well as new or additional discharges of effluent from pre-existing industries. All industries would be required to obtain a consent for any discharge from the State Board. The Act introduced explicit penalties of imprisonment, fine or both for: (i) discharging of effluent without permission; (ii) withholding of information when requested to supply it; (iii) failing to comply with the requests of pollution control officers in the case of emergency measures or Magistrates of at least Metropolitan or Judicial standing in other cases or requests from the Central Government; (iv) failing to provide information either upon request, or through negligence in reporting; (v) knowingly causing or permitting pollution of a water course; (vi) establishing a new industry without obtaining permission from the State Board; or (vii) failing to apply for a consent to discharge effluent after a period of notice that they should do so. Enhanced penalties were intended for subsequent convictions. A specific section of the Act made every person in charge of, or responsible to, a company personally responsible for a polluting act, as well as the company itself, unless the offence was committed without his or her knowledge, or if he or she exercised due diligence. If the offence was committed with the consent or connivance of any other, they also would be liable. The Act also explicitly excluded Civil Courts from having any jurisdiction over any suit or proceeding that arose form the Act, although later rulings clarified this as being in terms of removing the right of Civil Courts to remove the jurisdiction of authorities constituted under the Act, rather than removing the right of the Civil Court to entertain a suit or proceeding that restrains a defendant from causing pollution.

This Act was extended by the Water (Prevention and Control of Pollution) Cess Act (1977) to address continued problems with funding both the Central and State Pollution Control Boards. Under the Water Act (1974), Central and State Governments were required to provide necessary Board funding. State financial circumstances made this impracticable, and the new Act introduced a cess levy on local authorities, which were entrusted with the duty of supplying water, and on certain industries based upon the volume of water consumed.

The second major piece of legislation affecting water pollution was the Environment (Protection) Act (1986). This introduced general legislation for environmental protection, to recognise that some polluting industries had escaped the specifics of the 1974 Act. This extended automatically to the whole of India (rather than requiring approval on a state-by-state basis). It increased the penalties applicable to pollution offences to imprisonment for up to 5 years, an initial fine of one lakh (100 000) rupees, with additional fines of 5000Rs per day when the polluter continues to contravene regulations after the date of the first conviction. It also introduced the general principle that any member of the public could make a complaint under this Act, rather than requiring prosecution to be instigated by a Pollution Control Board.

Faced with problems with the State Pollution Control Boards, the Water (Prevention and Control of Pollution) Amendment Act (1988) allowed the Central Board to exercise the powers and perform the functions of the State Board in specific situations, such as when a State Board fails to act and comply with the directions issued by the Central Board. It also extended some of the remit of the Environmental Protection Act to those specific polluting activities covered by the Water Act (1974). This included allowing Boards to limit the period of a discharge consent to allow sufficient time for the Board to ensure that discharges were meeting consents, increasing penal provisions, and allowing the general principle that any member of the public could make a complaint under the Act. It also empowered Pollution Boards to close offending industries directly.

In addition to these general Acts, a number of specific rulings have been important in shaping the development of pollution regulation. Most importantly, these have clarified definitions of liability, extending them beyond their traditional legal definitions. A 1987 ruling (*M.C. Mehta* and *Union of India*) explicitly recognised the inappropriateness of English law for determining liability, arguing that *Rylands* and *Fletcher*, which advocated strict liability, was not appropriate for a country facing ever-increasing amounts of hazardous or inherently dangerous industries. This ruling argued that if an enterprise involved in a hazardous or inherently dangerous activity caused harm to anyone on account of an accident in the operation of the industry, the enterprise should be strictly and absolutely liable, without exception, and that the magnitude of any compensation provided should reflect the magnitude and capacity of the enterprise, in order to act as a deterrent for the future. The same ruling established the personal liability of the Chairman and the Managing Director of a polluting company, even where they did not know that the polluting offence was being committed, except where pollution was due to an Act of God or sabotage.

In 1988 the Supreme Court ruled that all new industries should be refused operating licences unless adequate provision had been made for the treatment of effluents before discharge, reinforcing provisions implicit in the Water Act (1974). More importantly, it clarified the Environmental Protection Act (1986) and Water Amendment Act (1988) in allowing any member of the public to move the Supreme Court to enforce the statutory provisions of these Acts, beyond situations where the petitioner was experiencing suffering directly. This argued that in situations where pollution was a public nuisance, then it would not be reasonable to expect any particular person to take proceedings to stop the pollution, as distinct from the community at large, and petition could be made under the banner of public interest litigation. A petitioner became entitled to move the Supreme Court to enforce the statutory provisions which impose duties upon the Boards under the Water Act. The Court also removed the remit that allowed High Courts to grant orders of stay on criminal proceedings when Pollution Control Boards initiated proceedings to prosecute industrialists or others polluting rivers, and specified that in the extraordinary granting of an order, this should be for only a short period.

This brief summary of water pollution legislation allows a number of conclusions. First, the legal basis for pollution control is exceptionally well developed. Although it is a hybrid of US and UK systems (Dwivedi and Khator, 1995), it places emphasis upon strict regulation with enforcement through the legal process. In many cases, either the law, or rulings on the law, go well beyond US and UK legislation, allowing third parties to bring prosecutions and having a much more stringent application of absolute liability. Second, central to the pollution control system are a set of Pollution Control Boards whose remit is defined by central legislation, often explicitly, but whose responsibility is to local State Government in the first instance. Such regulatory agencies are central to the effectiveness of any pollution control legislation, and their role and success will be conditioned by both their positioning within wider institutional structures, and the wider societal processes that will engage with the practice of the agency. The problem of ensuring effective agency performance is a problem returned to below.

FITTING THE LID: PUTTING POLLUTION REGULATIONS INTO PRACTICE

Despite the existence of seemingly powerful laws to protect the water environment, one of the main characteristics of India's water environment is continued pollution, and a failure of proper enforcement of pollution laws. There are a number of reasons for this: the history of pollution control; levels of pollution awareness; economic imperatives; and the role of the legal process.

The History of Pollution Control

It has been argued that in the presence of continuing or repetitive deviance, a "state of affairs" rather than an "act" (Hawkins, 1984), it is inevitable that sanctions-based approaches to pollution control are inappropriate, because they require a history of pollution control legislation and associated regulatory culture. In societies that have witnessed little pollution regulation, pollution is unlikely to be an exception to some general rule. In MDCs, the development of pollution has been incremental, in most cases over significant periods of time, allowing polluting industries to develop progressively the necessary knowledge of the technology required to curb the emission of harmful substances, and the acceptance that they should conform to imposed sanctions. Although LDCs such as India may have some history of pollution control, sometimes dating back to colonial rule, most industrial expansion, and hence environmental degradation, has been more recent. The development of pollution control policy is therefore taking place in a different context from that in MDCs, even though the approaches developed may draw upon MDC experience. For instance, Bowinder and Arvind (1989) note that the Ganga Action Plan in India has been prepared on the same lines as that developed for the River Thames in the UK, raising critical questions about the use of approaches informed by "Western" knowledge, out of the context for which they were intended. Whilst the scientific knowledge may be perfectly transportable, associated technology may not be, and this is further undermined by the assumption that strong and appropriate administrative systems exist for regulatory enforcement.

The lack of experience in pollution control undermines a sanctions-based approach for three reasons. First, if the development of pollution control is not incremental, sanctions-based approaches will be undermined by the lack of local knowledge of the causes and consequences of pollution problems and requisite mitigation measures. Polluters have no choice but to be non-compliant if they do not have such knowledge: their activities have become deviant through legislative development, without the provision of clear guidance as to why this activity is deviant, and what needs to be done to make it less so. This is not to argue that polluters should be seen as benign actors who have no choice other than to pollute. Indeed they may have the necessary knowledge, but still judge the economic costs of conforming to regulation to be so great as to warrant continued pollution. In essence, they fail to act as citizens in exploiting the environment as a free good (a pollution sink) in a similar manner to the loss of traditional indigenous systems for management of a common resource.

However, as Bradbaart (1995) found for water polluting industries in Indonesia, most polluters did not know what to do to become compliant, and were not aware of the economic costs *and benefits* of doing so. The reasons for continued deviance, and whether or not these are related to choice or constraint, need further investigation. Second, in situations where pollution is a state of affairs, rather than a set of discrete acts, enforcement of sanctions will result in the generation of large numbers of pollution cases which have to be dealt with by criminal or civil proceedings, both of which require time and money. Third, sanctions-based approaches rely upon the clear identification of a defendant, whose responsibility can be proved beyond doubt. However, detection is not readily assisted by the large number of potential polluters and the possible spatial extent of polluting activity. Monitoring large rivers, and the large number of potential polluters, could be helped by automated monitoring of environmental quality where that is possible. However, as was illustrated by the experience of the National Rivers Authority (NRA) in the UK, prosecutions may only be attempted if the source of pollution can be identified and the persons responsible for causing the event are known (NRA, 1994). In practice, the NRA relies extensively upon public observation of polluting incidents, as attributing cause requires rapid action. In many LDCs, there is neither the history of environmental monitoring, nor the necessary level of public awareness or concern, for detection to be possible; and these are issues that can be explained by both levels of pollution awareness and economic imperatives.

Education and Awareness

Following UK experience, enforcement of environmental regulations may require a public which is actively willing to engage in bringing deviant activities to the attention of regulatory agencies. People must be aware of the benefits of pollution control, and this causes one to emphasise the need to recognise different interpretations of pollution. The cultural categories that people use to classify, understand, and so become involved with their environment are important, not only because such knowledge has been acquired over long periods of time, but also because the concept of participation, which is successful for environmental management, has to be built on an understanding and mutual respect of such cultural systems (Brockenshal *et al.*, 1980; McNeely and Pitt, 1985; Richards, 1985; Redclift, 1987). This must recognise that science can generate information, but also that the information that is generated, the way in which this information is interpreted, and the significance attached to each piece of information, may vary according to particular social settings.

This in part reflects the fact that what is perceived as hazardous is, at least in part, societally determined. Most importantly, different levels of understanding of what constitutes environmental damage will reflect both education, and the interaction of that education with other forms of knowledge. For instance, the traditional definition of pollution in terms of some scientifically informed environmental standard may contrast markedly with social and religious understandings of what is and what is not polluted. Gunningham (1974, p. 20) notes '... pollution is a condition which departs from some standards of purity ... a social concept, not a scientific one...". For instance, for the millions of Hindus who bathe in the Ganga every day, the technically polluted waters are pure and cleansing, the living waters of the River of Heaven, such that the river, even when actually dirty, never is in the minds of those who believe (Ahmed, 1991). As Eck (1982, p. 216) notes "At question here ... is not really the purity of the Ganges, but the cultural understanding of what it means for something to be pure or impure, clean or dirty...". Whilst it may be possible for some to reconcile the "knowledge" that the river is polluted and the "knowledge" that it still has pure or healing properties, for many this subtle distinction may be lost, and the latter takes precedent over the former.

Economic Imperatives

A third reason for weak enforcement of pollution law stems from the close links that exist between environmental degradation and economic development, particularly at the local scale, where victims may stand to lose much from enforcement of pollution regulations. Local priorities in LDCs focus upon economic development, following from the need to tackle widespread poverty and improve people's living standards. As a result, environmental concerns in MDCs can become inverted in LDCs, where the environment becomes contested because its exploitation creates economic value, and not because it is valued for its amenity or aesthetic qualities. This inversion itself has origins in structural inequalities that exist between MDCs and LDCs as part of the global politico-economic system (Brundtland, 1987; Redclift, 1987). Concerns about economic development take precedence over the adverse environmental degradation consequent from such development, even if environmental degradation undermines both local development in general, and even longer-term economic development in particular (*cf.* Baker, 1990). If regulations are developed that attempt to prohibit a particular polluting activity, without some attempt to recognise the economic importance of that activity, both for the polluter and for some of the polluted whose livelihood may depend upon the polluting activity itself, then deviance is likely to continue. The continued failure of tanners in the region of Kanpur in Uttar Pradesh, India, to change their polluting activities, despite repeated orders by the Indian Supreme Court and evidence of the consequences of tannery pollution for local people, is in part an illustration of this. Similarly, Dasgupta (1997) notes how lead smelters in Calcutta contained a large number of small polluting units, many of whom could not afford pollution-reducing technology. Economic processes therefore condition and control regulatory effectiveness and the introduction of an economic dimension into pollution regulation increases the complexity of its enforcement in particular local contexts.

Both the levels of pollution awareness and economic imperatives imply that different levels of pollution are acceptable (not through choice, but rather by constraint) to different people in different places at different times. In this context, therefore, what may be regarded as morally wrong in one society may not necessarily be so in another, and pollution control can only be effective if it is tied to the ambitions and needs of particular polluters and victims. As Baker (1990, p. 46) "... The law, in short, is not a management instrument, it is a device to protect against abuse where alternatives exist...".

The Role of the Legal Process

One of the key assumptions of any regulatory approach is that the legal process is capable of dealing with deviant activity in a manner that does result in environmental protection when, for example, a regulatory agency takes a case to Court. Although Table 14.1 is reassuring in that it suggests that 80% of pollution cases in India that actually reach court result in a ruling in the Pollution Control Boards' favour, a very large number of cases do not reach the court, and it is likely that an even larger number of pollution cases go undetected. Furthermore, there are fundamental reasons against relying upon legal processes, even when they do rule in the state's favour under criminal law or the plaintiff's favour under civil law.

First, once decision making is in the hands of a legal process, it becomes distant from those who are experiencing the adverse effects of pollution, such that awareness of the full extent of damage is reduced, and it becomes less likely that compensation or remuneration for damage will be sufficient to allow the problem to be redressed. For instance, in India, even where cases reach the Supreme Court (SC) and decisions are reached in favour of the plaintive, rulings have not necessarily provided adequate resources to redress damage. Sharma and Banerji (1996) describe how a ruling of the SC on 13 February 1996 was in favour of the plaintiffs, ordering that five units of an acid polluting plant in Bichri, Rajasthan, should be closed down. However, despite

Table 14.1 *The status of water pollution prosecution cases raised under violations of the Water Act up until 30 September 1989, by State Pollution Control Boards, and consequent from changes introduced by the Environment Protection Act (1986). Modified from Bowinder and Arind (1989)*

State Pollution Control Board	Total cases	Action completed	Ruled in favour of the Board	Ruled against the Board
Andhra Pradesh	–	–	–	–
Assam	–	–	–	–
Bihar	144	3	3	–
Goa	–	–	–	–
Gujarat	1201	234	148	86
Haryana	301	120	105	15
Himachal Pradesh	64	21	13	8
Karnataka	79	6	4	2
Kerala	36	16	13	3
Maharashtra	319	22	13	9
Madhya Pradesh	101	2	–	2
Orissa	43	2	1	1
Punjab	351	75	31	44
Rajasthan	169	22	13	9
Uttar Pradesh	160	66	66	–
West Bengal	14	–	–	–
Central Pollution Control Board	192	124	122	2
Total	**3593**	**746**	**553**	**193**

significant environmental degradation as a result of the polluting activity (e.g. pollution levels in both surface and groundwaters that were so high that no natural water sources could be used in the area), the SC awarded no exemplary damages to local people. The SC ruling only applied to immovable assets. The company had filed for bankruptcy, assets were insufficient to meet liabilities, and it is thought that the polluter had moved its assets to a different state where it was operating an identical acid plant. The ruling also placed the claims of secured creditors over unsecured ones, such that any monies generated from the sale of immovable assets would not accrue to the local Pollution Control Board, the body responsible for attempting to restore water supplies. In this example, although a successful prosecution has been made under criminal law, the victims are not guaranteed adequate compensation. This is not a new problem: as the NRA (1994) notes for England and Wales, the size of fines imposed by Crown Courts on polluters does not reflect the severity of the pollution alone, however that is determined, but also reflects the defendant's means or ability to pay. Again, this is a reflection of the inevitable separation of legal activity from experience of the real impact of a pollution problem.

Second, relying upon a sanctions-based approach implies a commitment to cure rather than prevention. Deviance is dealt with after the event: once a case gets to a court, the damage has been done, and in many situations may not be easily reversed. For instance, in the case reported by Sharma and Banerji (1996), there remains dissent over whether or not polluted water can be treated to ensure it is safe for either animal or human consumption. This might not necessarily apply if court action resulted in recommendations or effects that extend beyond the pollution case in question, to prevent pollution occurring in the future. Recommendations might include reform in the relevant pollution control bodies to allow a more preventative approach. In the case reported by Sharma and Banerji (1996), to prevent similar situations occurring in the future, it is necessary for agencies involved at the state and local levels to be more proactive in their engagement with pollution problems. However, even though some of the blame for a failure to take action lay with the Ministry of Environment and Forests, neither the Central Pollution Control Board, the Rajasthan Pollution Control Board nor the Rajasthan State Government, all of whom could be described as carrying out their duties negligently, were criticised by the Supreme Court or required to modify the way in which they were engaging with pollution control. Similarly, the fact that large numbers of polluters continue to engage in deviant activity after a series of landmark decisions in the Indian Supreme Court, both in favour of Pollution Control Boards and plaintiffs, suggests that the effects of particular rulings are not sufficient to encourage polluters to conform to regulations.

Third, as Bowinder and Arvind (1989) note, fines cannot be extended to include exemplary damages under criminal law, and the small size of fines has meant that polluting industries have tended to find it more economically advantageous not to comply and to continue to risk paying penalties. To obtain exemplary damages, and thereby to achieve ultimately better environmental protection, requires a clear signal to polluters by imposition of large damages, and it is necessary to refer to civil law. However, for countries like India this remains problematic as it requires victims to sue polluters responsible for causing them damage. This requires both finance and time, which is likely to be lacking for those sectors of the population who are most likely to be the victims of pollution. In some cases in India, progress has been made through the activities of philanthropic advocates, willing to act on behalf of perceived victims (e.g. M.C. Mehta), but the number of lawyers with the requisite knowledge, education and willingness to act on behalf of the poor remains small. Further, dependence upon civil law requires the development of tort and associated case law, the latter being lacking in India (Bowinder and Arvind, 1989).

Fourthly, both criminal and civil law tend to draw extensively upon precedent, which causes problems given the need to identify conformity between cases. Most pollution events will have caused various levels of damage (dependent upon the magnitude and type of polluting activity and the assimilative capacity of the receiving environment), will have affected the local population in different ways, and will have different clean-up costs. It remains unclear as to the extent to which earlier cases can provide reliable guidance in new and probably very different pollution cases.

One of the assumptions of a more tort-based approach is that it is only when polluting firms pay for the damage that they have caused that they will seek technological alternatives that can reduce discharge at source. Reduction at source (internalising the externalities) is clearly a key means of preventing environmental damage, rather than being committed to curative strategies. However, it also assumes that polluters know what to do to become compliant and are aware of the benefits that might come from compliance. Some approaches to pollution regulation in LDCs have been referred to as ineffective (e.g. Bernstein, 1955; Zwick and Benstock, 1971;

Freeman and Haveman, 1973; Gunningham, 1974) as a result of weak enforcement. Yet these criticisms fail to consider why polluters fail to comply and the extent to which misinformation, misinterpretation and misunderstanding encourage deviant behaviour.

Space and Spatially Delimited Enforcement

The final reason for the failure of a sanctions-based approach relates to the problem of achieving full spatial jurisdiction. Detection and enforcement appear to be spatially irregular, and enforcement tends to be localised, with the inevitable consequence that polluters simply shift production in response to regulation, an inevitable consequence if a polluter perceives that there is no other viable alternative. For instant, in the example from Rajasthan referred to above, the polluter ordered to close under the SC ruling of 13 February 1996 simply relocated to a different state (Gujarat), continuing to undertake exactly the same type of polluting activity (Sharma and Banerji, 1996). The continued shift of pollution from urban to rural areas is one supported by a Supreme Court ruling in 1986 (in *M.C. Mehta* and *Union of India*), and the Government of India is actively and currently encouraging polluting industry to move out of urban areas to improve urban environments, and in some situations, to encourage rural development (Anon. 1996). This tendency to shift pollution, often away from the environments most able to deal with it (for historical reasons, larger cities tend to be located on rivers with greater discharges and hence assimilative capacities), is a highly political process, and one that exists when polluters are not aware of the costs and benefits of complying to regulation in an existing location. It follows that an approach in which agencies work more co-operatively with deviant polluters, and with other agencies, might begin to address this problem.

LIFTING THE LID: WOULD A MORE COMPLIANCE-BASED APPROACH WORK?

It follows from the discussion above that the mere existence of regulations, however stringent, is insufficient to ensure their adoption. Scholz (1984a) argues that a compliance-based approach has two major advantages. First, a more co-operative strategy will require fewer enforcement resources, with detection and prosecution focusing upon bad firms. Second, it may actually encourage polluters to adopt good practice, by providing an incentive for voluntary compliance, which would excuse the polluter from some of the time and inconvenience that comes from dealing with repeated requests for inspection by the regulatory agency.

Empirical research (e.g. Bradbaart, 1995) supports the idea that a compliance-based strategy can result in higher levels of environmental protection, as polluters are convinced of the importance and worth of better pollution control, and are better educated as to the means by which this may be achieved. Bradbaart found for wastewater treatment in Indonesia, that continual non-compliance could be explained by: (i) excessive costs of compliance; (ii) lack of popular pressure forcing industries to comply; (iii) disputes over the legitimacy of environmental laws; (iv) poor communication channels between governments and agencies, such that environmental laws were seen as rhetoric rather than reality; and (v) poorly developed local expertise in pollution control. Bradbaart found that (ii) to (v) were the most important reasons, and clearly imply that pollution control needs to be informed by local pollution control officials, who are willing to work with polluters, to a stage where pollution becomes a deviant act (checked by regulations) rather than a general state of affairs. Similarly, Dasgupta (1997) found that many small polluting industries needed guidance from regulatory authorities so that they would become able to meet established environmental standards, particularly in the lead-smelting sector where a large number of both owners and workers belonged to a migrant group with little or no formal education, and where there was a low level of understanding of pollution standards and what should be done to meet them. Thus, more effective enforcement of pollution regulations is only likely to occur once pollution is no longer a state-of-affairs, but a small, finite number of deviant acts, such that the polluter may be readily identified, and regulatory cultures have developed such that pollution is really seen as a criminal act.

However, choice of a compliance-based strategy raises important questions, both of justice and of practicality. Underpinning the development of water pollution law is the definition of acts of environmental violation as criminal offences. Acceptance of a compliance-based approach implicitly recognises that water pollution is not strictly a criminal offence, placing less emphasis upon punishment and more upon deterrence. In practice this is a distinction of degree rather than of kind, as in all criminal offences deterrence will be one aim of punishment. Once guilt has been established, a judge can use some discretion about the nature and severity of punishment. A compliance-based approach could be viewed in the same way, but with legal procedures side-stepped: in the presence of deviant activity, regulatory agencies determine the optimal means of achieving compliance for each individual polluter. This clearly may help to achieve the aim of pollution regulation, in ensuring better environmental protection, but it undermines both the regulatory principle that polluting the environment is a criminal act, and may undermine the economic principle that polluters ought to be made to pay for the damage that they do (e.g. OECD, 1975), according to the way in which compliance is achieved.

There is some support for viewing environmental offences as different in kind from those of traditional criminal acts. Even in societies with a history of pollution control, there may be a general reluctance to regard a breach of environmental regulation as morally reprehensible. The conduct being addressed is widely regarded as morally neutral (e.g. Kadish, 1963; Ball and Friedman, 1963; Yoder, 1978) in contrast with the behaviour that is normally the concern of traditional criminal law. This moral ambivalence results in the difficulties of detecting deviant activity described above and is one of the reasons why pollution continues in a sanctions-based system. Ambivalence arises: (i) because many victims are remote in time from deviant activity (for instance, it may take some time for the effects of pollution to be seen in ill-health); and (ii) because as a result of this remoteness, and the differing degrees to which exposure to pollution causes harm, victims are dispersed, so reducing the ability to identify a criminal, and to form cohesive pressure groups. Water pollution has a much stronger association with general loss of environmental value, whether that be in use, amenity, aesthetic or ecological terms – concerns that may be described as luxuries in LDC contexts. This is further reinforced by the practice of civil courts, which place great emphasis upon establishing the level of damage a particular individual has experienced. Even in criminal justice hearings in India, with a constitution that is explicit in making environmental destruction a criminal offence, the associated criminal justice system is largely oriented around identification of human victims, associated with a legacy of criminal justice adopted from the Western world. Choice of a compliance-based approach raises awkward questions of justice: whether or not pollution should be viewed as a criminal activity, and if so whether or not it is just for regulatory agencies to circumvent a criminal justice procedure, even if the latter is oriented around a system with a different definition of victim.

TIGHTENING THE LID: INSTITUTIONS AND THE "ADMINISTRATIVE TRAP"

The second context for evaluation of a compliance-based approach is the institutional one. Implementation of both sanctions-based and compliance-based approaches requires regulatory agencies who, regardless of approach, have similar responsibilities. These agencies: (i) must ensure that environmental goals are met (whether in terms of effluent discharge or qualities of the receiving environment); (ii) may be required to determine appropriate levels of regulation, if legislation does not do so, or if other statutory bodies are not appointed to do so; (iii) must monitor industries and receiving environments to detect non-compliance; and (iv) must take appropriate steps when goals are not being met. It is under (iv) that the two distinct styles of pollution regulation emerge. Under a sanctions-based approach, a polluting industry which is non-compliant would be referred directly to the legal system. Under a compliance-based approach, the agency would engage in a more consultative approach, attempting to negotiate with polluters to achieve maximum compliance, and only resorting to prosecution when polluters continually fail to comply with the requests of the agency. Thus, regulatory agencies will exert a critical conditioning process upon the effectiveness of legislation, both in terms of the regulatory styles that they choose and the extent to which they pursue them effectively.

Evidence for the failings of regulatory agencies in India is widespread. Table 14.1 shows that although 3593 cases of violation of environmental regulations were reported by State Pollution Control Boards, only 20% had been completed, partly because of the time taken to complete a case, but more especially because many of the cases were dropped by the Pollution Control Boards before completion. There remains a general unwillingness to implement sanctions-based strategies by pollution control authorities, and to follow prosecution cases through to their completion.

Consideration of the experience of pollution control suggests two broad reasons why this is the case: (i) the position of regulatory agencies within broader political and economic processes; and (ii) overlapping and unclear responsibilities. First, in LDCs, regulatory agencies are faced with the difficulty of preserving a fragile balance between the interests of economic activity on the one hand and public welfare on the other. Both pollution control staff and industrialists have been shown to hold the view that the design and enforcement of regulation affects not only environmental quality, but also industrial productivity and profitability, and hence the conditions under which manufacturers compete in the domestic and international market-place (Hawkins, 1984). One of the consequences of pollution regulation has been the development of differential environmental standards, both within countries and between countries, that could have both direct and indirect effects upon industrial profitability, and hence industrial location. Competition to survive may be so intense, that there is little ability to invest in capital-intensive pollution treatment facilities (Dwivedi and Khator, 1995).

Given the potential economic effects of pollution regulation, it is hardly surprising that those agencies responsible for regulation find their legitimacy questioned by opposing public constituencies (Dwivedi and Khator, 1995), although organisational self-preservation requires them to appear to be acting, achieving and in control (Hawkins, 1984). The effect of this situation is the development of manipulative relationships, often corrupt, between polluters and regulators (Dwivedi and Khator, 1995), what Dwivedi (1989) calls "black administrations". For instance, the activities of Pollution Control Boards in India are often either ignored, or manipulated by the concerns that they are trying to regulate: the Supreme Court discovered in August 1995 that certain highly polluting industries in Uttar Pradesh had been allowed by the Uttar Pradesh Pollution Control Board (UPPCB) to function with relaxed environmental standards. Some (e.g. Selznick, 1966) see developments such as these as an inevitable consequence of regulation wherein an agency is co-opted by those it seeks to regulate, incorporating and reflecting their concerns into its decision making in the interests of stability and self-preservation. Bernstein (1955, p. 83) notes how "...the mores, attitudes, and thinking of those regulated come to prevail in the approach and thinking..." of many regulatory officials. The economic imperative may become particularly problematic when the polluters themselves are part of the state. Abdel-Motaal (1996) found in Egypt that only 12% of public sector violations of pollution regulations had been dealt with under available legislation, as compared with 64% in the private sector. They note that this is because it is difficult to apply the law to the public sector as the regulator and the polluter is the state in both cases.

Second, research has shown how the activities of regulating agencies may be compounded by the existence of a multiplicity of pollution controlling agencies involved in the regulation process (Central, State, regional and municipal), sometimes working at cross-purposes. Dwivedi and Khator (1995) note that the Indian Central Government is generally responsible for making policy and associated legal mechanisms, but the State Governments are responsible for implementing them, through Pollution Control Boards. However, the Pollution Control Boards are also responsible to the Central Ministry of Environment and Forests, in terms of meeting national environmental standards. The State Government may view a project as necessary for economic sustainability, encouraging the State Pollution Control Board to be "reasonable" in its judgement, whereas the Central Ministry views the pollution problem from a national standpoint, and may pressure the state environment agency to apply universal standards. The consequence is a complicated balancing act, which effectively reduces the ability of the regulatory agencies to implement pollution control, even though it must be seen to be making efforts to do so. For instance, the UPPCB has the power to close any polluting industry, but it has rarely exercised this power, preferring to distance itself from the act of regulation by referring the matter to the Indian Supreme Court, rather than directly engaging with the polluting industry.

The examples illustrate how regulatory agencies do not exist independently of those that they seek to regulate and as a result cannot be relied upon to put pollution control into practice, and raise questions about the assumption that a more compliance-based approach will be any better than a sanctions-based approach. Can agencies engage in more collaborative approaches to pollution control, without their goals being subsumed into the desires of polluters to avoid regulation, and compliance becoming complicity?

At first assessment, it would be tempting to argue that a more "negotiated" treatment of polluters would only encourage them to be more lax, as good firms take advantage of an apparently weak regulator. However, this needs to be evaluated in both theoretical and empirical terms. Theoretically, Scholz (1984a,b) used a simple game theory model to establish the conditions in which a compliance-based approach should be adopted, whilst showing that the net social benefits are higher by minimising enforcement and compliance costs through co-operation between agency and firm. Such an approach is useful as it recognises the inherent uncertainties that the regulated perceives in terms of regulator behaviour, and the regulator perceives in terms of behaviour of the regulated. At any one time (Table 14.2), Scholz (1984a) shows that the cheapest option for the firm is to evade a co-operative agency, due to the reduced probability of detection of deviant activity when an agency is taking a more co-operative view of the firm. The most expensive option is to comply with a deterring agency, because the firm is being denied the cost saving that comes from illegal emission, whilst experiencing the extra administrative costs of dealing with a hostile regulator.

From the agency perspective, taking a stringent attitude, and making a firm comply will achieve the pollution reduction. However, the agency knows that if the firm realises it is being stringent, it is much cheaper to evade than to comply, which achieves less pollution reduction. The agency achieves more pollution reduction if it chooses a more co-operative stance *but only if* the firm complies. This is where the dilemma (essentially a "prisoner's dilemma") arises as the optimal situation is where the firm complies and the agency is co-operative, but both sides know that they are worst off if

Table 14.2   *The enforcement dilemma. Modified from Scholz (1984a)*

		Agency approach	
		Co-operative emphasis	Deterrence emphasis
Industry strategy	Comply	Good firm voluntarily complying with a co-operative agency. Cost to firm: $3 million Pollution reduction: 400 tonnes	Good firm harassed by a stringent agency. Cost to firm: $5 million Pollution reduction: 425 tonnes
	Evade	Bad firm tempted to evade a co-operative agency. Cost to firm: $2 million Pollution reduction 160 tonnes	Bad firm confronted by a stringent agency. Cost to firm: $4 million Pollution reduction: 320 tonnes

the other cheats. If the firm evades a co-operative agency, pollution reduction will be reduced. If the agency takes a stringent stance, and the firm continues to comply, the firm pays much more in pollution control.

This dilemma is extended by Scholz (1984b) to the situation where the game is played through time, recognising that decisions by both firm and agency will be conditioned by the history of decision making. For instance, a co-operative agency that detects a firm that has been tempted to evade it, is more likely to view the firm as "bad" and to switch to a more stringent deterrent strategy both to punish the firm, and to ensure better environmental protection. Likewise, a good firm, complying with an apparently co-operative agency, might be more likely to choose evasion, if it discovered that the agency was actually acting in a stringent manner. Scholz (1984b) shows that the game works in both the agency and the firm's favour if: (i) the agency pursues a strategy with both the vengefulness and forgivingness of tit-for-tat; and (ii) the benefits of compliance are clear to the firm such that the value of future co-operation to the firm is great enough to compensate for the current temptation of the firm to cheat.

The results of this analysis suggest that an approach that simply enforces sanctions using an ever more stringent approach is likely to fail. Scholz's analysis shows that the effect of simply enforcing the law ever more stringently will be to increase the cost of pollution regulation to firms when they are compliant, causing them instead to choose evasion, and to take advantage of the low probability of detection. In India, the low probability of detection, combined with low levels of completion of cases by a legal process (Table 14.1) and a relatively low cost of fines, reinforces the low cost of evasion, even if an agency is stringent. It implies that greater detection is important, but as noted above, the extent and complexity of water pollution problems, and the unwillingness and inability of the public and the media to report pollution incidents, means that increases in detection are difficult to achieve, regardless of how stringent an agency might become.

However, it is important to realise that the real-world behaviour of agencies and firms is much more complicated than a model like this can accommodate. Indeed, Scholz (1984b) notes that actual co-operation will inevitably fall short of what is potentially achievable because of the problems of detection, because firms fail to recognise the advantages of co-operation, and because of institutional instability. All of these preclude the development of a policy environment in which the rules of the game become well established. To this must be added the problem of "black administration" referred to above, in which the agency and the firm share common interests which are not necessarily those of better environmental protection. This model assumes that the agency's sole goal is reduction of pollution to an acceptable level. Corruption will clearly challenge this assumption. Similarly, the model assumes much about the sociological behaviour of both agency and firm. For instance, it assumes that the administrative system is amenable to negotiation. However, as Dwivedi and Khator (1995) note, this is generally not true of India's civil service which instead tends to be excessively rigid and ineffectively bureaucratic. Dwivedi (1989) argues that much of the reason for this can be traced back to the colonial origins of the Indian civil service, which imitated the British system of supposedly objective, neutral civil servants, who relied upon doing things by the book. Dwivedi argues that this, combined with a growing influence from corrupt political influences, has in turn allowed corruption to influence certain aspects of the Indian civil service, until co-operative agency behaviour is unlikely because of a rigid adherence to rules, which are only broken by corruption. This, of course, undermines the conclusions suggested by Scholz's model. At Dalmau, on the River Ganges, downstream from the city of Kanpur, Uttar Pradesh, a water monitoring station, according to local officials, is continually non-operational. However, the Head of the Local Pollution Control Board claims that it is operating. Such a discrepancy partly reflects a hierarchical society in which there are established systems of authority which determine the way in which knowledge is obtained and evaluated. Although this may seem corrupt from a Western perspective, an alternative interpretation would argue that this is inevitable and even acceptable in an under-funded civil service, in which individual performance can make a material difference to one's standard of living, and where personal priorities arise in part from a history of institutional development that has made corruption acceptable. Thus, issues over how pollution regulations are best enforced can only be evaluated in particular places, and research of this kind is critical in different socio-cultural contexts.

CONCLUSIONS

Bottom-up, compliance-based approaches may be more effective because they recognise: (i) that the pressures of economic development require that maximum use is made of local environmental assimilative capacity; (ii) that societal processes will determine the ease and effectiveness of enforcement; and (iii) that those without access to, and control of, resources, technology or knowledge, must be included in the pollution control process if pollution control programmes are at least ensuring social benefits and justice (Ahmed, 1991). However, this analysis has attempted to show that such successes remain dependent upon the effectiveness of the institutions responsible for enforcing environmental regulations, for detecting pollution and identifying criminal behaviour, for monitoring environmental quality and even for involving local people, through a better education, as to the importance of pollution control. For instance, Dwivedi (1994) suggests that institutions must consult and work with polluting industries (*cf.* Bradbaart, 1995), must be allowed discretionary powers, and must be given clearer jurisdiction and responsibility. This conclusion follows, whether or not a sanctions-based or compliance-based approach is adopted. Similarly, the legal system must be developed, because of the special nature of pollution in general, and water pollution in particular, perhaps to include a separate environmental tribunal which allows accused polluters reasonable opportunity to make their case quickly (Dwivedi, 1994). Real environmental improvement requires continuing institutional evaluation and re-evaluation combined with improvements in the legal basis for pollution control.

   However, the issue of water pollution in Less Developed Countries must be considered much more broadly than through a simple comparison of different styles of regulatory enforcement. Whilst it is tempting to argue that all that is needed to improve environmental protection is a specific environmental court capable of handling the volume and nature of pollution cases, and the development of more effective or "sustainable" administrations, such arguments fail to address the real reasons why regulations are not working. In a society where there remains considerable pressure to increase economic output, any judgement on the philosophical basis of environmental law must recognise the need for regulation to deal with abuse of environmental externalities resulting from market failure, so that the water environment is protected. The problem is goal-oriented, and it is more important that polluters comply rather than that criminals are prosecuted. This does not just require compliance-based systems to work more effectively. Pollution, and the responses of polluters, those who suffer pollution and regulatory agencies are implicitly bound up in processes which condition and create a political and economic landscape in which regulatory enforcement operates, whether sanctions-oriented or compliance-oriented. For instance, the economic imperatives which help explain a certain ambivalence to pollution control are themselves conditioned by larger scale inequalities in the global economic system. Not only do these emphasise economic priorities, but they are one of the reasons why polluting industries find it continually advantageous to locate in Less Developed Countries, whether this is the result of direct and intentional policy making (e.g. Ives, 1985) on the behalf of multinational companies, or simple and subtle shifts in terms of trade that add the savings from not complying to environmental regulation to low labour and raw material costs. Likewise, just as more effective enforcement of environmental regulation in some countries improves the water environment of those countries, this may well be to the detriment of other countries where a reduced ability to enforce regulation allows the continued existence, and even expansion, of similarly polluting industries elsewhere. These observations suggest that in addition to considering the legal and institutional frameworks that surround pollution control, thought must be given to the way in which these processes are conditioned by broader political and economic processes, to avoid choosing solutions that are bound to fail because they treat symptom rather than cause.

ACKNOWLEDGEMENTS

Nicky Padfield (Fitzwilliam College, University of Cambridge) provided critical guidance as to the legal basis of water pollution control. S.S. was in receipt of a Nehru Scholarship and S.W. is in receipt of a Cambridge Commonwealth Trust Scholarship.

REFERENCES

Abdel-Motaal, D. 1996. *Pollution control in the River Nile in Egypt and the Polluter Pays Principle*. Thesis submitted to the University of Cambridge in partial fulfilment of the conditions for award of the Master of Philosophy Degree in Environment and Development.

Ahmed, S. 1991. *Questioning Participation: Culture and Power in Water Pollution Control: the Implementation of the Ganga Action Plan at Varanasi*. Unpublished PhD thesis, University of Cambridge.

Anon. 1996. *The Times of India*, 13 December 1996.

Baker, R. 1990. Institutional innovation, development and environmental management: an 'administrative trap' revisited. *Public Administration and Development*, **9**, 29–47.

Ball, H.V. and Friedman, L.M. 1963. The use of criminal sanctions in the enforcement of economic regulations: a sociological view. *Stanford Law Review*, **17**, 197–223.

Banerji, R. 1996. Dead fish tales. *Down to Earth*, **5**(17), 27–33.

Banerji, R. and Martin, M. 1997. Courting green. *Down to Earth*, **5**, 27–35.

Bernstein, M.H. 1955. *Regulating Business by Independent Commission*. Princeton University Press, New Jersey.

Black, D.J. 1976. *The Behaviour of Law*. Academic Press, New York.

Bowinder, B. and Arvind, S.S. 1989. Environmental impact: environmental regulations and litigation in India. *Project Appraisal*, **4**, 182–196.

Bradbaart, O., 1995. Regulatory strategies and rational polluters – industrial waste-water control in Indonesia, 1982–1992. *Third World Planning Review*, **17** 439–458.

Brockenshal, D., Warren, D.M. and Werner, O. (Eds), 1980. *Indigenous Knowledge Systems and Development*. University Press of America, Washington, DC.

Brundtland, H. 1987. *Our Common Future*. Oxford University Press, Oxford.

Caldwell, L.K. 1970. Authority and responsibility for environmental administration. *The Annals of the American Academy of Political and Social Science*, **389**, 107–115.

Caldwell, L.K. 1987. The contextual basis for environmental decision-making: assumptions are predeterminants of choice. *Environmental Professional*, **9**, 302–308.

Chakraborty, S. 1995. Green justice: up in smoke? *Down to Earth*, **4**(9), 19–23.

Dasgupta, N. 1997. Greening small recycling firms: the case of lead-smelting units in Calcutta. *Environment and Urbanization*, **9**, 289–305.

Dwivedi, O.P. 1989. Administrative heritage, morality and challenges in the sub-continent since the British Raj. *Public Administration and Development*, **9**, 245–252.

Dwivedi, O.P. 1994. *Development Administration: from underdevelopment to sustainable development*. Macmillan, Basingstoke, 161pp.

Dwivedi, O.P. and Khator, R. 1995. India's environmental policy: Programs and politics. In Dwivedi, O.P and Vajpeyi, D.K. (Eds), *Environmental Policies in the Third World: A comparative analysis*. Mansell, London, 47–70.

Eck, D.C. 1982. *Banaras – City of Light*. Alfred A. Knopt, New York.

Eisenberg, M.A. 1976. Private ordering through negotiation: dispute-settlement and rule-making. *Harvard Law Review*, **89**, 637–681.

Freeman, A.M. III and Haveman, R.H. 1973. Clean rhetoric and dirty water. In Enthoven, A.C. and Freeman, A.M. III (Eds), *Pollution, Resources and the Environment*. Norton, New York.

Gunningham, N. 1974. *Pollution, Social Interest and the Law*. Martin Robertson, London.

Hart, H.C. 1988. Political leadership in India: dimensions and limits. In Kohli, A. (Ed.), *India's Democracy: An Analysis of Changing State-Society Relations*. Cambridge University Press, Cambridge.

Hawkins, K. 1984. *Environment and Enforcement*. Oxford University Press, Oxford.

Ives, J.D. (Ed.) 1985. *The Export of Hazard: transnational corporations and environmental control issues*. Routledge, Boston.

Kadish, S.H. 1963. Some observations on the use of criminal sanctions in enforcing economic regulations. *University of Chicago Law Review*, **30**, 423–449.

McNeely, J. and Pitt, D. (Eds) 1985. *Culture and Conservation: The Human Dimension of Environmental Planning*. Croom Helm, London.

Mitchell, B. 1971. Behavioural aspects of water management: a paradigm and a case study. *Environment and Behaviour*, **3**, 135–153.

NRA (National Rivers Authority). 1994. *Water Pollution Incidents in England and Wales – 1993*. HMSO, London.

O'Riordan, T. 1981. *Environmentalism*, 2nd edition. Pion, London.

OECD (Organisation for Economic Co-operation and Development). 1975. *The Polluter Pays Principle: Definition, analysis and implementation*. OECD, Paris.

Padfield, N.M. 1995. Clean water and muddy causation: Is causation a question of law or fact, or just a way of allocating blame? *The Criminal Law Review* (September), 683.

Redclift, M. 1987. *Sustainable Development: Exploring the Contradictions*. Methuen, London.

Reiss, A.J. Jn and Biderman, A.D. 1980. *Data Sources on White-Collar Law-Breaking*. National

Institute of Justice, Washington, DC.

Richards, P. 1985. *Indigenous Agricultural Revolution: Ecology and Food Production in West Africa.* Longman, London.

Sagoff, M. 1988. *The Economy of the Earth: Philosophy, Law and the Environment.* Cambridge University Press, Cambridge.

Scholz, J.T. 1984a. Voluntary compliance and regulatory enforcement. *Law and Policy*, **6**, 385–404.

Scholz, J.T. 1984b. Deterence, co-operation and the ecology of regulatory enforcement. *Law and Society Review*, **18**, 179–224.

Selznick, P. 1966. *TVA and the Grass Roots: A Study in the Sociology of Formal Organisation*, revised edition. Harper and Row, New York.

Sewell, W.R.D. 1971. Environmental perceptions and attitudes of engineers and public health officials. *Environment and Behaviour*, **3**, 23–59.

Sharma, A. 1996. When a river weeps. *Down to Earth*, **4**(22), 27–31.

Sharma, A. and Banerji, R. 1996. The Blind Court. *Down to Earth*, **4**, 27–34.

Yoder, S.A. 1978. Criminal sanctions for corporate illegality. *Journal of Criminal Law and Criminology*, **69**, 40–58.

Zwick, D. and Benstock, M. 1971. *Water Wasteland.* Grossman, New York.

15

Towards Integrated Management of Rural Drainage Basins with Particular Reference to Water Quality Issues

T.P. Burt

INTRODUCTION

The British countryside has long been the subject of conflict between individuals and organisations who place different, and sometimes contradictory, demands on the land. Since World War II, this disharmony has intensified, the main focus of attention being the clash between modern farming practices and the need for wholesome drinking water. More recently other demands have emerged: a widespread interest in nature conservation, and a greater demand for recreational opportunities in the countryside, especially from those who increasingly live locally in towns and villages.

In modern Europe the ready supply of wholesome drinking water from public supplies is taken for granted. So is the plentiful supply of food. This report is largely about the problems which can arise when these two basic needs conflict. This is how the House of Lords' 1989 report *Nitrate in Water* began. The report considered a proposal for EU legislation (since adopted as Directive 91/676) to protect fresh and marine waters from pollution from nitrate from diffuse sources. The quotation reminds us of the varied functions of rural land: farming is not the only "use" of land, and where there is discord, as between food production and water supply, compromises will be needed.

Water Quality: Processes and Policy. Edited by Stephen T. Trudgill, Des E. Walling and Bruce W. Webb.
© 1999 John Wiley & Sons Ltd.

PROBLEMS OF MODERN FARMING

Generations of farmers have adjusted their use of land to the combined influence of market forces, available technology and the inherent nature of the land. Land use for most of the twentieth century has reflected an emphasis on production, particularly since World War II when the need for self-sufficiency in food production became clear. Farming has become steadily more intensive as farmers have met the challenge of government policy and social trends fuelling an increasingly capital-led industry. Expensive heavy machinery, use of fertiliser, land drainage, frequency of ploughing – all reflect the demands for increasing arable output. Grassland has become more productive too, with reclamation of rough grazing, reseeding of old pasture, and more use of fertiliser and manure.

As noted above, there is now growing recognition of the multiple use of land; farming must compete with these other "uses". An important function of rural land is as a catchment area for rivers, reservoirs and aquifers. Inevitably conflicts have arisen as modern agriculture has impacted upon water supplies, especially in relation to the pollution load of water draining from farmland. There is also increased interest in farmland from many sections of the general public: some demand better access to the countryside, others have an interest in nature conservation, while for many more, farmland simply provides the pleasing backdrop for a drive in the country. Given the high level of public subsidy for farming over recent decades, it is perhaps not surprising that the public have demanded more say in how the nation's land is managed. At the same time, modern farmers, inevitably driven by the need for profit, have found it more difficult to sustain their position as stewards of the countryside.

The impact of modern farming is seen in two ways: on and off the farm. On the farm, concerns relate to soil degradation and to the loss of habitats for wildlife. With the clearance of woodland, drainage of riverside land and grubbing up of hedgerows, areas of high conservation interest, such as Sites of Special Scientific Interest, have become more isolated in the landscape. Recent schemes to assist farmers in tree planting and in the establishment of riparian buffer zones should provide enhanced conservation interest and a more pleasant and varied landscape. Other initiatives have sought to sustain traditional countryside crafts, such as dry stone walling and hedging, to retain unimproved hay meadows, and to prevent further drainage of low-lying ground. By offering incentives to farmers to manage their land in an environmentally beneficial way, there is now positive linkage between Common Agricultural Policy (CAP) reforms and the management of rural land. The agri-environment package (2078/92) introduced at the same time as the MacSharry price support reforms was very limited in scope, though its share of the CAP budget has now risen to 4%, proving that transfers across budget headings within the CAP are at least possible (Whitby *et al.*, 1996).

Soil erosion is traditionally thought of as a feature of semi-arid lands, but many British soils are now threatened. This is not so much a result of hostile climate as of a lack of awareness by farmers and policy makers, and continuing financial incentives to grow arable crops on erosion-prone land. In the past, erosion by wind was perceived to be the main problem but today the main risk is known to be erosion by water, particularly on the intensively farmed lighter soils of the lowlands (Evans, 1990). Farmers have been encouraged by generous subsidies to grow cereals on quite unsuitable land. Soil conservation is possible, but this is rarely an attractive option unless itself subject to grant aid. At present soil conservation schemes are voluntary and there is no means of targetting erosion-prone sites. Compaction by heavy machinery, loss of organic matter, a trend towards autumn-sown crops (leaving ground bare over the winter) and larger fields all contribute to the problem (Boardman, 1990). The issue is not so much one of reduced crop yields through loss of nutritious topsoil, though this might become important before too long, but of off-farm impacts. Flooding of property by soil-laden water has become a contentious issue in many areas of northwestern Europe (Boardman *et al.*, 1994), river channels and lakes can become clogged with sediment, and roads may be inundated. The eroded sediments carry with them pollutants such as phosphate (Sharpley *et al.*, 1994) and pesticides (Harris and Forster, 1997); these can severely affect the health of aquatic ecosystems downstream. There is also significant erosion of peat soils in the uplands, where over-grazing has contributed to the problem (Anderson *et al.*, 1997); sedimentation of reservoirs is a particular problem in these areas (Labadz *et al.*, 1991).

Most concern has been expressed about the impact of nitrate leached from farmland (Burt *et al.* 1993). Unlike phosphate (where the equation is more finely balanced; see Withers, 1993), farming is the main source of nitrate in river and groundwater. It is over-simplistic, however, to blame the problem on fertiliser use alone; the nitrate issue has many contributing sources including the increased frequency of ploughing, land drainage, new crops such as oil seed rape which require high fertiliser applications, and, in some areas, higher stocking densities. Attention is now being given to prevention of nitrate leaching at source through a combination of good farming practice and schemes which provide financial compensation in return for significant changes in land use and management practices (Burt and Haycock, 1993). Given greater legislative control of "point" discharges from sewage treatment works, diffuse (or "non-point") losses of nitrogen and phosphorus from farmland have become the main source of nutrient enrichment in our rivers, lakes and coastal waters. Losses both of soluble and sediment-bound pesticides remain an important cause for concern too.

Agricultural policies of successive governments (both UK and EU) have been a major cause of the countryside being over-exploited by intensive farming. It is debatable whether their objectives (which include increased production, a fair standard of living for the agricultural community, and available supplies at reasonable prices) have been met; what is clear is that the cost, including environmental degradation, has been high (Whitby *et al.*, 1996). In the twenty-first century we may anticipate more cross-compliance between financial support for farming and the need to maximise environmental protection. Soil conservation *sensu lato* must become an important element of such policies as farmers seek to sustain soil productivity and minimise off-farm impacts. In this way, productive, sustainable farming may co-exist alongside other rural land uses, providing clean water, better access, improved habitats and a diverse attractive landscape.

TOWARDS CONSENSUS

As a result of the "Earth Summit" held at Rio de Janeiro by the United Nations Conference on Environment and Development in 1992, the action plan Agenda 21 was drawn up, in order to make progress in achieving sustainable development. Sustainable management of water resources forms an important element of Agenda 21:

> Water resources must be planned and managed in an integrated and holistic way... By the year 2000 all states should have national action programmes for water management, based on catchment basins or sub-basins, and efficient water-use programmes. These could include integration of water resource planning, land use planning and other development and conservation activities...

Gardiner (1994) points out that management of natural resources is hindered by incomplete scientific knowledge. This uncertainty has encouraged the application of the precautionary principle and a move towards prevention as a more sustainable solution. According to the World Conservation Union and others (IUCN/UNEP/WWF, 1991), sustainable development involves "improving the quality of human life while living within the carrying capacity of supporting ecosystems". In relation to water quality, this requires integrated management of the entire drainage basin, rather than piecemeal "cures" at specific locations. Pinay *et al.* (1990) argue that unidirectional flow promotes mixing of materials from upstream towards downstream, giving a dependence of downstream sites on headwater zones. These linkages are underlined in unifying ideas in aquatic ecology such as the river continuum concept (Vannote *et al.*, 1980) and nutrient spiralling (Webster, 1975). Burt and Haycock (1992) have stressed the need for integrated catchment planning in relation to the nitrate issue.

Gardiner (1994) has argued that the move from "cure" towards "prevention" as a more sustainable solution to water quality management requires the environmental argument to be made stronger in the decision-making process. Involvement of all stakeholders, i.e. those individuals or organisations with a legitimate interest in the outcome of the decisions being made, would facilitate this strengthening. Gardiner notes that this implies progression from "public consultation" (by government agencies) to "community participation". Robins (1995) argues for a similar sequence, moving from "planning of people" through "planning for people" and "planning with people" to "planning by people". "Planning with people" views the community as having a legitimate and rightful role in articulating social objectives on a routine basis. Robins notes that "planning by people" is rarely possible in the literal sense but that, nevertheless, "planning with people" and "planning by people" provide a platform for compromise and trade-offs between rival interests and subsequently opportunities for real social choices (Cassells and Valentine, 1988). Bass (1995) notes the success of participatory strategies: the advantages of stronger partnerships between stakeholders include more involvement of community and private sector interests with less domination by government agencies, greater public debate and understanding of issues, a broader foundation of ideas, skills and local knowledge, a better basis for reaching consensus on trade-offs, improved accountability and political credibility and, perhaps most importantly, greater commitment to implementation on all sides. Robins (1995) mentions the following benefits of stakeholder participation in policy making: legit-imacy, representation, empowerment, open communication and accessibility to infor-mation. The aim of participatory planning is not to avoid conflict but to enable stakeholders to be able to voice their concerns, ensuring that suitable structures are in place to deal with multiple interests positively (Robins, 1995).

The need to manage water quality in drainage basins sustainably requires a new approach in both time and space:

- seeking long-term solutions as well as quick fixes;
- accepting a measure of uncertainty and risk;
- dealing with a multiplicity of stakeholders with diverse values and interests;
- recognising an interdependency of environmental, economic and social issues;
- above all, considering the entire catchment area in a holistic manner.

In Australia, the "Landcare" movement has achieved a quiet revolution (Campbell, 1995) in land and water management involving farmers, government employees and the local community. This innovative approach may suggest a way forward for integrated catchment management elsewhere.

LANDCARE

The history of European settlement in Australia is very short but even by the 1930s there was concern about soil erosion. Over the last fifty years a wide variety of stubborn problems have arisen: severe soil erosion and gullying, salinisation, waterlog-ging, eutrophication, invasion of non-native plants (e.g. camphor laurel and the thorny shrub *Mimosa pigra*), and animals (most notably rabbits). Many of these difficulties affect not just one but several adjacent properties. Landcare was born in the early 1980s in northern Victoria where neighbouring landholders formed the Warrenbayne–Boho Landcare group to co-ordinate action to tackle transboundary degradation problems (Burton, 1988, in Robins, 1995). This approach was formalised as the Victorian Landcare Program by the State Government and the Victorian Farmers Federation in 1986. As an indication of progress, 68 registered Landcare groups existed in Victoria alone by 1990, involving 3180 landowners, expanding to 94 by early 1992 with more than 5500 landholders. The benefits of a participatory approach to land and water management were subsequently recognised and embraced at the national level. In 1989 the Federal Government, encouraged by the National Farmers Federation and the Australian Conservation Foundation, launched a National Land-care Program with a budget of A$340 million (Robins, 1995). Despite tough economic conditions in rural communities, the explosive growth of the Landcare movement has continued, with over 2000 groups by 1994 involving one-third of Australian farming families (Campbell, 1995). Campbell ponders why the number of Landcare groups has grown so quickly. People form groups because they recognise that groups are more effective than individuals; they realise that certain problems like erosion and weeds require co-operation; and they recognise the heightened sense of community which results.

Robins (1995) notes that participatory approaches to natural resource management combining integrated catchment management and Landcare are increasingly being incorporated into both Australian State and Federal law. For example, the Victorian Catchment and Land Protection Act 1994 establishes and empowers catchment-based co-ordination groups across the entire State comprising key agency and community representatives, including Landcare group members. Such legislation moves away from the orthodox "command and control" towards a more "education and persuasion" approach to natural resource management (Robins, 1995). Robins summarises by noting that Australians now demand participatory policy-making frameworks and governments have come to realise the value of stakeholder involvement.

Campbell (1995) notes that the Australian Landcare movement is highly differentiated so that it is difficult to describe a "typical" group except in broad terms as a voluntary group of (usually rural) people working together to develop more sustainable systems of land and water management. Both Robins and Campbell identify the catalytic role of government agency staff and local community leaders. Government staff now act as facilitators, much less concerned with transferring technical information than with providing opportunities for farmers and others to open their minds to new possibilities and to consider new approaches to looking at their situation. Robins recalls her early experience as part of an inter-agency funded project team working in the Wimmera catchment: none of the team members had any formal training in facilitation or conflict resolution. However, when working on the Yarra catchment only a few years later, she received relevant training, alongside community leaders and other key agency staff. Robins (1995) provides detailed case studies of both the Wimmera and Yarra catchment management schemes, emphasising the value of "grass-roots" participatory planning approaches in rural areas.

CATCHMENT MANAGEMENT IN THE UK

In England and Wales rather little has been achieved in bringing together those with a legitimate interest in water quality management in rural catchments; however, recent signs are more promising. The National Rivers Authority (NRA) was created by the Water Act 1989 to be the principal agent responsible for safeguarding and improving the water environment in England and Wales. Given its motto "Guardian of the Water Environment" and in line with the UK Government's commitment to sustainable development, the NRA set about preparing a collection of 164 catchment management plans (CMPs) for all the river basins in England and Wales. Each plan was to achieve the following objectives (Newson, 1996): outline a future for the basin which will allow it to develop its full environmental potential; identify conflicts between the NRA's own duties and operations and find ways to resolve them; and engage the general public (and more particularly their local government planners) in debate about the compatible uses of land. The CMP programme was a major undertaking and certainly an important step forward. However, unlike Landcare the approach was distinctly "top down" with the NRA identified as the lead agency for most purposes. While other key participants were identified (e.g. County Councils, local Naturalist Trusts, riparian landowners), it remained unclear how implementation of the plans was to proceed, particularly since in most instances the NRA did not own the catchment areas for which it was producing plans.

The Environment Act 1995 incorporated the NRA into the newly formed Environment Agency (EA), which began operation on 1 April 1996. CMPs metamorphosed into Local Environment Agency Plans (LEAPs). If its *Draft Water Quality Strategy*, published on 11 December 1996, is a guide, the EA can be expected to be much more proactive in terms of partnership and stakeholder participation. The overall aim of the Strategy is to achieve a continuing and overall improvement in water quality. This is to be achieved, *inter alia*, by developing a preventative approach by considering all potential sources of pollution, controlling and preventing pollution caused by discharges from both point and diffuse sources, ensuring that dischargers pay the costs of the consequences of their pollution, and by working in partnership with others. The Strategy identifies two categories of principal stakeholders: "customers" – the public at large, the water industry, other national trade and industry groups (e.g. National Farmers Union), interest groups (e.g. Friends of the Earth); and "partners" – Government Agencies (Department of the Environment, Ministry of Agriculture, Fisheries and Food – MAFF), local authorities, environmental regulators (most particularly the European Union), "quangos" (e.g. English Nature), and universities and research councils. The Strategy recognises "the need for collaboration and co-operation with other bodies, regulated organisations and the public to ensure that necessary action is progressed". Of course, the EA must exercise certain duties within the legislative framework. Even so, it is to be hoped that it will vigorously pursue a policy of participatory policy making which will give other stakeholders real representation in the planning process *and* will be prepared to compromise and trade-off where appropriate. Consultation is not the same as participation and it remains to be seen how far the EA will move in recognising the legitimate interests of others. Participation remains "mere tokenism" if government officials do not take the inputs of other groups seriously, and a key constraint will arise if the EA (at all levels in its organisation) is not genuinely committed to community involvement. In any case, the EA can achieve little in terms of non-point pollution without the co-operation of others, most notably the farming community.

In Scotland, where a different legal system and organisational structure pertains, there is no statutory requirement to promote integrated catchment management (ICM). Nevertheless, recent changes to the structure of the water industry in Scotland (notably the creation of the Scottish Environment Protection Agency and the new Water Authorities) have created an atmosphere within which ICM might develop. Werritty (1995) concludes that there is a strong case for introducing ICM into Scotland, favouring an "intermediate" model in which government agencies target a strategic issue (e.g. eutrophication) which is not just of local significance and seek to promote resolution locally via concentrated action by a variety of stakeholders. In this way it might be possible to reconcile a national policy for river management with action targetted upon issues that both generate public concern and will be owned by local communities.

SOME EXAMPLES OF ICM IN THE UK

Notwithstanding the absence of any statutory requirement to promote ICM in the UK, a number of schemes have been implemented in recent years (Figure 15.1). All focus on an issue of local concern and involve partnerships of one sort or another. Some of these schemes are reviewed briefly here.

Water of Leith (Scotland)

The Water of Leith Integrated Environment Action Plan arose from an initiative by the Lothians Branch of the Scottish Wildlife Trust (SWT) in 1989 to promote the protection and management of wildlife along the Water of Leith. Funding obtained by SWT allowed a River Valleys Officer to be appointed. A management plan published in 1993 addressed enhanced water quality, the conservation of wildlife and improved amenity. The plan established the Water of Leith Action Group, a partnership of several agencies including District Councils, Forth River Purification Board, Forth Ports plc, Scottish Natural Heritage (SNH) and SWT. Werritty (1995) comments that this is an excellent example of a "bottom-up" initiative which has succeeded in bringing into a single forum all the key stakeholders needed to revitalise this specific urban watercourse.

Loch Leven

Loch Leven is a highly eutrophic loch fed by small streams draining fertile agricultural land in the central lowlands of Scotland. It has an important brown trout fishery and its shores include a National Nature Reserve; it is an important site for migratory and wintering wildfowl. In 1995 a Steering Group was established to develop a CMP for promoting ICM at Loch Leven. This Group comprises staff from SNH, Forth River Purification Board, the Scottish Agricultural College and the local District Council. Among the tasks to be undertaken are the reduction of phosphorus loadings through changes in land use and farming practice, including creation of buffer strips. Although the CMP will be funded and managed by the Steering Group, extensive consultation with farmers and other stakeholders is envisaged (Werritty, 1995).

West Galloway Rivers Trust

This was created in 1989 by four district salmon fishery boards who saw advantage in an alliance of interests to address the question of falling salmon and sea-trout catches since the early 1960s. Acidification of surface waters due to acid atmospheric deposition, probably exacerbated by the impact of coniferous plantation in many parts of the four catchments, is generally blamed for the decline. Though funded by the salmon fishery boards, the Trust has been supported by a wide range of organisations including the Regional Council, Scottish Office Environment Department, SNH and

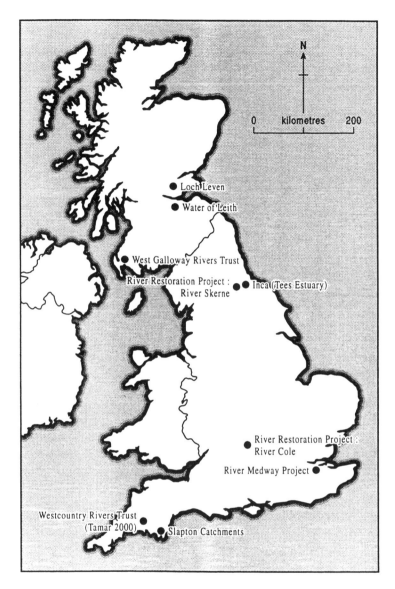

Figure 15.1 *Approximate location of catchment management schemes mentioned in the text*

the Forestry Commission (Stephen, 1994). Given its non-statutory status, the Trust is necessarily dependent on partnership with landowners and statutory agencies for its eventual success (Werritty, 1995). Current work includes electro-fishing surveys and water chemistry analysis to establish a clearer picture of the problem, and the development of a Geographical Information System (liaison with the Geography Department at Durham University) to aid catchment management.

Cleveland Industry Nature Conservation Association (INCA)

This partnership of industrial companies, Local Authorities, English Nature, the EA and local conservation interests aims to ensure an improving environment in a heavily industrialised area. Though not explicitly set up with ICM in mind, INCA has focused on the River Tees catchment, most notably its estuary, and structurally has much in common with the Australian examples discussed earlier. INCA seeks to promote constructive relationships with regulators and authorities and positive environmental developments by local industries; it acts as an advisory service for its private sector members and provides a communication channel with the public. INCA has also funded research projects, including some on water quality.

Medway River Project

This was established in 1988 to manage and enhance the landscape, wildlife and recreational amenity of the Medway Navigation, and to promote community aware-ness of the river environment. Its projects have attracted an impressive amount of voluntary support from the local community and provide a focus for much environ-mental education in the district. The Project has been supported by industrial sponsor-ship and by grant aiding from both local and national schemes. Projects have focused on improvements to public access and bank restoration, to prevent erosion and restore riverside habitat (Smith, 1994).

Westcountry Rivers Trust

This Trust was formed to "conserve, maintain and improve the natural beauty and ecological integrity of rivers and streams in the Devon, Cornwall and West Somerset countryside, the restoration of sympathetic flow regimes of pure water being central to the recovery of river corridor biodiversity". The Trust seeks to involve the local community and to work in liaison with other organisations. It aims to promote the enjoyment and wise use of water resources for recreation, to encourage use of river basins for environmental education, to provide practical advice, and to support relevant research. Its Tamar 2000 project provides a clear example of ICM and places great emphasis on forging links with various individuals and organisations: riparian landowners, farmers and their representatives, environmental and wildlife groups, resource management agencies, all levels of government, education and research centres, and other companies and individuals as appropriate. The project aims to restore to a healthier condition, a catchment which has suffered environmental degra-dation as a result of harmful land use practices (relating to dairy farming and live-stock rearing) in recent decades. The project is seeking funding from relevant UK and EU funds in order to be able to offer technical advice and financial support to farmers.

River Restoration Project

The River Restoration Project has received major funding under the EU LIFE programme in order to progress two demonstration projects in river channel restoration. On the River Cole, a tributary of the upper Thames, two lengths of channel which had been straightened in the 1960s for land drainage purposes have had meandering channels engineered. In one case the original channel was recreated and in the other a completely new set of meanders was designed. There has been close liaison with the local landowners, the National Trust, and with the EA. Future plans include construction of a wetland to buffer discharge from an under-drained clay subcatchment and implementation of floodplain buffer zones along the new meandering channel. On the River Skerne in Darlington, a new meandering channel has also been created and other riparian enhancements effected along this heavily urbanised stretch of river. In both cases, the aim is to recreate instream and riparian ecological processes, improve biodiversity, and so restore the nutrient recycling capacity of the rivers.

Slapton Catchments

Slapton Ley is a Site of Special Scientific Interest and was created a National Nature Reserve (NNR) in 1993. It is the largest natural body of freshwater in southwest England ("Ley" is a local word for lake) and a site of international importance for flora and fauna. There have been concerns over eutrophication and sedimentation of the Ley since the 1960s: intensive mixed farming has led to increased soil erosion and nutrient export from the Slapton catchments in recent decades (Burt *et al.*, 1996). The catchments lie within an Area of Outstanding Natural Beauty and the eastern boundary of the Ley forms part of the Heritage Coastline. Tourism is therefore an important element both within the NNR and beyond and there is great need to maintain landscape quality. Robins formulated a plan for management of the Slapton catchments in 1995, using her experience of ICM and Landcare in Australia (see above). Several meetings were held involving key stakeholders: English Nature, South Hams District Council, local farmers and the National Farmers Union, NRA, MAFF and the nature reserve owners (Whitley Wildlife Trust) and managers (Field Studies Council). At present the plan has stalled, lacking funding for a project manager and support from some agencies. The Landcare experience emphasises the need for a local organiser to catalyse others into action and it may be that this element is currently lacking too. It remains to be seen whether an Australian initiative can be successfully transplanted into rural Devon! It is worth adding that the Slapton catchments have been chosen as one of the pilot sites for MAFF's Water Fringe Habitat Scheme. Following modest take-up, more flexible regulations have been introduced in an attempt to encourage wider participation (Tytherleigh, 1997). At present, the NNR Management Committee may be the best hope of moving towards a fully fledged ICM scheme. This Committee comprises the NNR landlord, tenant and English Nature, together with observers from local parish councils and the Heritage Coastline Service, giving it an unusually wide basis already. Since the EA is about to produce a

LEAP for that part of south Devon which includes the Slapton catchments, this may be a good time to invite the EA to join the Committee. Without proper attention to catchment processes, management of a lake system like Slapton Ley remains highly problematic.

CONCLUSIONS

If the Environment Agency's eventual policy on water quality looks anything like its draft Strategy, it will have adopted a strong commitment to partnership and stake-holder participation. This would be most welcome and would tie in well with the emerging opportunities for a more radical approach to agri-environmental policy via CAP reforms (Whitby *et al.*, 1996). The time appears ripe for greater flexibility in rural land management and there is now greater encouragement for wider participation in the planning process via Agenda 21 and other initiatives. In the UK, most attention has focused so far on the intensively farmed lowlands, but ICM may be an equally relevant approach in the uplands where familiar issues such as erosion, landscape appearance, access and wildlife conservation are also the subject of much discussion (Anderson *et al.*, 1997; Burt *et al*, 1997). Thus far, "bottom-up" ICM schemes in the UK have been on a modest scale. Landcare provides a seductive template for ICM; it remains to be seen whether such a revolutionary approach proves too radical for the traditionally conservative inhabitants of rural Britain or for similarly cautious agencies and organisations. Assuming that the Environment Agency maintains its commitment to partnership, it may well be that an approach similar to that suggested by Werritty (1995) for Scotland will work best in England and Wales too: targeting by the EA of a strategic issue which is of widespread significance, then seeking to promote consensus locally by encouraging full participation of all interested parties – in other words a sensible mixture of "top-down" and "bottom up" approaches.

NOTE IN PROOF

Since writing this chapter, a LEAP for the area which includes the Slapton Catchments has been published by the EA. A voluntary action group is proposed for the Slapton Catchments, a welcome outcome of earlier efforts.

REFERENCES

Anderson, P., Tallis, J.H. and Yalden, D. 1997. *Restoring Moorland*. Peak District Moorland Management Project, Phase III report, Peak National Park, Bakewell, Derbyshire.
Bass, S. 1995. *Participation in Policy Processes*. IIED, London.
Boardman, J. 1990. Soil erosion on the South Downs: a review. In Boardman, J., Foster, I.D.L. and Dearing, J. (Eds), *Soil Erosion on Agricultural Land*, Wiley, Chichester, 107–118.
Boardman. J., Ligneau, L., de Roo, A. and Vandaele, K. 1994. Flooding of property by runoff from agricultural land in northwestern Europe. *Geomorphology*, **10**, 183–196.

Burt, T.P. and Haycock, N.E. 1992. Catchment planning and the nitrate issue: a UK perspective. *Progress in Physical Geography*, **16**(4), 379–404.

Burt, T.P. and Haycock, N.E. 1993. Controlling losses of nitrate by changing land use. In Burt, T.P., Heathwaite, A.L. and Trudgill, S.T. (Eds), *Nitrate: Processes, Patterns and Management*. Wiley, Chichester, 341–367.

Burt, T.P., Heathwaite, A.L. and Trudgill, S.T. 1993. *Nitrate: Processes, Patterns and Management*. Wiley, Chichester.

Burt, T.P., Heathwaite, A.L. and Johnes, P.J. 1996. Stream water quality and nutrient export in the Slapton catchments. *Field Studies*, **8(4)**, 613–627.

Burt, T.P., Labadz, J.C. and Butcher, D.P. 1997. The hydrology and fluvial geomorphology of blanket peat: implications for integrated catchment management. Paper presented at *Mires Research Group Meeting*, Manchester, 10 April.

Campbell, C.A. 1995. Landcare: participative Australian approaches to inquiry and learning for sustainability. *Journal of Soil and Water Conservation*, (March–April), 125–131.

Cassells, D.S. and Valentine, P.S. 1988. From conflict to consensus: towards a framework for community control of the public forests and wildlands. *Australian Forestry*, **51**, 47–56.

Evans, R. 1990. Soil erosion: its impact on the English and Welsh landscape since woodland clearance. In Boardman, J., Foster, I.D.L. and Dearing, J. (Eds), *Soil Erosion on Agricultural Land*. Wiley, Chichester, 231–254.

Gardiner, J. 1994. *A sustainable water environment: the operational challenge*. Paper presented at the Institution of Civil Engineers, 8 December.

Harris, G.L. and Forster, A. 1997. Pesticide contamination of surface waters – the potential role of buffer zones. In Haycock, N.E., Burt, T.P., Goulding, K.W.T. and Pinay, G. (Eds), *Buffer Zones: their Processes and Potential in Water Protection*. Quest Environmental, Harpenden, 62–69.

IUCN/UNEP/WWF (The World Conservation Union, United Nations Environment Programme, World Wildlife Fund for Nature) 1991. *Caring for the Earth: a Strategy for Sustainable Living*. Gland, Switzerland.

Labadz, J.C., Burt, T.P. and Potter, A.W.R. 1991. Sediment yields and delivery in the blanket peat moorlands of the southern Pennines. *Earth Surface Processes and Landforms*, **16**, 265–271.

Newson, M.D. 1996. Catchment plans: a new geographical resource. *Geography Review*, **9**(3), 17–24.

Pinay, G., Decamps, H., Chauvet, E. and Fustec, E. 1990. Functions of ecotones in fluvial systems. In Naiman, R.J. and Decamps, H. (Eds), *The Ecology and Management of Aquatic-Terrestrial Ecotones*. Parthenon (UNESCO), Paris, 141–169.

Robins, L. 1995. *Structures and processes for community participation in integrated catchment management: strategies for the Wimmera and Yarra catchments, Victoria, Australia*. Unpublished MSc thesis, University of Oxford.

Sharpley, A.N., Chapra, S.C., Wederpohl, R., Sims, J.T., Daniel, T.C. and Reddy, K.R. 1994. Managing agricultural phosphorus for the protection of surface waters: issues and options. *Journal of Environmental Quality*, **23**, 437–451.

Smith, B. 1994. *Medway River Project: Six Year Review 1988–94*. Medway River Project, Maidstone, Kent.

Stephen, A.B. 1994. *West Galloway Fisheries Trust: 5 year Review and Progress Report 1989–1994*. West Galloway Fisheries Trust, Newton Stewart, Scotland.

Tytherleigh, A. 1997. The establishment of buffer zones – the Habitat Scheme Water Fringe Option, UK. In Haycock, N.E., Burt, T.P., Goulding, K.W.T. and Pinay, G. (Eds), *Buffer Zones: their Processes and Potential in Water Protection*. Quest Environmental, Harpenden, 255–264.

Vannote, R.L., Minshall, G.W., Cummins, K.W., Sedall, J.R. and Cushing, C.E. 1980. The river continuum concept. *Canadian Journal of Fisheries and Aquatic Sciences*, **37**, 130–137.

Werritty, A. 1995. Integrated catchment management: a review and evaluation. *Scottish Natural Heritage Review*, **58**, 70pp.

Webster, J.R. 1975. *Analysis of potassium and calcium dynamics in stream ecosystems on three southern Appalachian watersheds of contrasting vegetation.* Unpublished PhD thesis, University of Georgia, Athens.

Whitby, M., Hodge, I., Lowe, P. and Saunders, C. 1996. Conservation options for CAP reform. *ECOS*, **17**(3/4), 46–54.

Withers, P.J.A. 1993. The significance of agriculture as a source of phosphorus transported to surface waters in the UK. Paper presented at Society of Chemical Industry symposium *Phosphorus and the Environment*, 19 October.

Author index

Subject index